Exploring Materials through Patent Information

Exploring Materials through Patent Information

David Segal
Abingdon, Oxfordshire, UK

THE QUEEN'S AWARDS
FOR ENTERPRISE:
INTERNATIONAL TRADE
2013

Print ISBN: 978-1-78262-112-6

A catalogue record for this book is available from the British Library

Published by The Royal Society of Chemistry,
Thomas Graham House, Science Park, Milton Road,
Cambridge CB4 0WF, UK

Registered Charity Number 207890

Visit our website at www.rsc.org/books

Printed and bound by CPI Group (UK) Ltd, Croydon, CR04YY

Dedicated to my mother and father

Preface

Rapid changes in technology are having an increasing impact on everyday life, particularly on communications, consumer products and healthcare. For example, mobile phone technology and social media networks have had a large effect on the way people communicate and exchange information. Changes in technology often involve the use of materials including metals, ceramics and polymers, which are used to fabricate integrated circuits, blood glucose monitors, three-dimensional or 3D printing of plastic prototypes and light-emitting diodes in displays. However, not all materials can be described easily as either metal, ceramic or polymer, for example, ionic liquids are salts that are liquid at room temperature. Some materials are combinations of metal, ceramic and polymer, such as ceramic fibre- or powder-reinforced metallic composites. The early decades of the 21st century are indeed an age of materials. Many people may not realise that the disciplines of physics, chemistry, materials science, mathematics and engineering underpin the technical features of inventions that are described in the worldwide patent literature and inventions are reflected by the changes in technology observed in everyday life from communications through to consumer goods and healthcare.

Anyone with a computer link can, in principle, access a vast quantity of information by carrying out searches on the Internet. However, my discussions with undergraduates, lecturers,

Exploring Materials through Patent Information
By David Segal
© David Segal, 2015
Published by the Royal Society of Chemistry, www.rsc.org

business people and members of the general public indicate a lack of awareness that patent literature is a primary source of scientific, technical and medical information and is publicly accessible from national patent offices. In my experience, many excellent textbooks use journal references to aid the reader, but make little use of patent literature and the information described in the literature.

This book is not an instruction manual for describing how to carry out patent searches. Neither is it a reference work on intellectual property, as I am not qualified to give formal legal advice on patenting issues. Comments I have made on patenting procedures represent practical guidelines for readers. Patent literature represents a world occupied by words and their meanings as defined by inventors and their patent attorneys. This is not a world inhabited by glossy illustrations and, for this reason, I have included drawings throughout the text for clarification, but no photographs. The world of patents can appear to be opaque to anyone entering it for the first time. The language can be arcane and the sentences can be very long, so that studying the documents is a time-consuming process.

I have attempted to show in this book that patent information described in the patent literature can be used to help explain developments in diverse areas of technology. A short monograph cannot be a substitute for an encyclopaedia, and I have selected a range of materials that have specific application or potential application for discussion in separate chapters. These chapters do not replace excellent review articles that may be found on specific materials and indeed this book can complement reviews that concentrate on references other than patents. The patent literature allows, in principle anyone to determine how a technology has developed over time. This development over time is particularly pertinent for three-dimensional printing, which is attracting much publicity in the public domain. However, the contents of this book do not represent a historical study and I have concentrated on interpretation of technical information in the patent literature in order to compile the chapters.

I hope that this monograph will encourage a younger generation of students, as well as interested readers to routinely consider patent information as an important source of scientific, technical and medical information alongside other sources of information.

David Segal
Abingdon, UK

About the Author

David Segal read natural sciences at Trinity Hall, Cambridge University and graduated in 1972. He obtained a MSc with commendation in surface chemistry and colloids at the University of Bristol and completed a PhD on foaming in lubricating oils at the same university. Post-doctoral work was carried out at Brunel University on the surface chemistry of copper phthalocyanine pigments. He has been employed in both the public and private sectors in the United Kingdom and throughout his working life has maintained an interest in materials chemistry and patent literature. David Segal wrote the book *Chemical synthesis of advanced ceramic materials*, which was published by Cambridge University Press in 1989. He is an author or co-author of over forty scientific papers and an inventor or co-inventor of over twenty-five patent families.

Symbols

A, B	Linkage groups
a.c.	Alternating current
AM	Additive manufacturing
a, x, y, z	Atomic fractions
AMOLED	Active matrix organic light-emitting displays
CAD	Computer-aided design
CCFL	Cold cathode fluorescent lamp
cDLP	Continuous digital light processing
CMOS	Complementary metal-oxide-semiconductor
cm	Centimetre
$^{\circ}$C	Degrees centigrade
d.c.	Direct current
DMLS	Direct metal laser sintering
E_e	Electron energies
E, F, G, X, Y	Terminal groups
FWHM	Full width at half-maximum
FDM	Fused deposition modeling
f_A, f_B	Mole fraction of polymers
FIB	Focused ion beams
$\varepsilon_{parallel}$	Dielectric constant parallel to long axis of molecule
$\varepsilon_{perpendicular}$	Dielectric constant perpendicular to long axis of molecule
γ	Surface tension of liquid
ΔG_{mix}	Free energy for mixing polymers

Exploring Materials through Patent Information
By David Segal
© David Segal, 2015
Published by the Royal Society of Chemistry, www.rsc.org

HOMO	Highest occupied molecular orbital
INID	Internationally agreed numbers for the identification of bibliographic data
k	Boltzmann constant
K	Degrees Kelvin
LCD	Liquid crystal display
LDV	Laser Doppler vibrometer
LED	Light-emitting diode
LUMO	Lowest unoccupied molecular orbital
M^I, M^{II}	Metal
M	Moles
MAPS	Microwave assisted processing and synthesis
MBE	Molecular beam epitaxy
mg	Milligram
MOCVD	Metalorganic chemical vapour deposition
mm	Millimetre
MRI	Magnetic resonance imaging
mW	Milliwatt
n, m	Degree of polymerisation for polymers A and B
nm	Nanometre
NTSC	U.S. National Television System Committee
OLED	Organic light-emitting diode
p	Pitch
Δp	Pressure difference
PHOLED	Phosphorescent organic light-emitting displays
PLED	Polymer light-emitting diode
%	Percentage
PMOLED	Passive matrix organic light-emitting displays
r_p	Pore radius
RIE	Reactive ion etching
SLM	Selective laser melting
SLS	Selective laser sintering
SMOLED	Small molecule light-emitting diode
SpO_2	Blood oxygen saturation levels
Θ_c	Contact angle
TFT	Thin film transistor
μm	Micrometre
T	Absolute temperature
UKIPO	United Kingdom Intellectual Property Office

USPTO	United States Patent and Trademark Office
V	Volts
W	Watts
WIPO	World Intellectual Property Organization
χ	Flory–Huggins interaction parameter

Contents

Exploring Materials through Patent Information
By David Segal
© David Segal, 2015
Published by the Royal Society of Chemistry, www.rsc.org

CHAPTER 1

Introduction: Exploring the World of Materials through Patent Information

1.1 INTRODUCTION

Developments in all aspects of science and technology are having an increasing impact on everyday life. In communications, social media networks have become popular across the world as a way of communicating and exchanging information between people. In healthcare, magnetic resonance imaging (MRI) and computerised tomography are routine diagnostic tools in hospitals. Stents, spinal implants, hip and knee replacements are used in surgical procedures, while self-testing for measuring blood glucose concentrations is widely used. Mobile phones, personal computers (whether desktop, laptop, notebook or tablet), games consoles, televisions, digital cameras, in-vehicle navigation systems and inkjet printers are examples of everyday consumer products and devices. Product development requires input from fundamental disciplines, namely physics, chemistry, materials science, mathematics and engineering. This input often relates to materials, including metals, ceramics or polymers, their fabrication and properties. However, not all materials can be described easily as either metal, ceramic or polymer; for

Exploring Materials through Patent Information
By David Segal
© David Segal, 2015
Published by the Royal Society of Chemistry, www.rsc.org

example, ionic liquids are salts that are liquid at room temperature. Some materials are combinations of metal, ceramic and polymer, thus we have ceramic fibre- or powder-reinforced metallic composites. The input from the scientific disciplines mentioned above may not be obvious to the general public, students, lecturers or anyone with business interests. Developments in lighting technology involving light-emitting diodes are spreading out to encompass applications such as televisions and displays in mobile phones. Additive manufacturing techniques that include three-dimensional (3D) printing have applications in diverse fields, medicine and aerospace, while advances in semiconductor fabrication are supported by fundamental studies in physics, chemistry and materials science. The early decades of the 21st century represent an age of materials.

Physics, chemistry, materials science, mathematics and engineering are the disciplines that underpin the features of inventions generally that are described in the worldwide patent literature. Anyone with an interest in science, technology and medicine may want to gain detailed information on particular topics by accessing relevant literature. Nowadays, access to vast amounts of information can, in principle, be achieved through a computer link to the Internet. However, computer users may be unaware that patent literature is kept at national patent offices and is publicly available. Patent literature is a primary source of information that allows monitoring of technological developments.

In this chapter, a range of materials relevant to technological change is summarised. However, this book is not an encyclopaedia and this range is not comprehensive. In later chapters selected topics with application to everyday life have been chosen in an attempt to show how useful patent information can be in understanding developments in underlying technologies. While the contents of these chapters may not be of direct interest to the activities of some readers, it is hoped that those readers will appreciate how useful patent information can be in following the development and advances in their own areas of interest.

A brief overview of patents is given in this chapter. It needs to be stressed that readers are advised to obtain advice from qualified practitioners such as patent attorneys, especially if they intend to prepare patent specifications.

1.2 MATERIALS UNDERPINNING TECHNOLOGICAL ADVANCES

Examples of materials that underpin technological advances are given here. Some of these materials are observable by the general public but many are hidden from view. Semiconductor devices such as transistors on integrated circuits are built-up on a high-purity silicon wafer or chip by using photolithography and etching combined with curable organic resins known as photoresists; curing is achieved on exposure to laser light. Transistors are produced by doping the silicon with trace elements, arsenic, boron or phosphorus, for example, which modify electron transport in the silicon. Layers of devices are separated by electrically insulating layers of glass or silica to prevent electrical interference and devices are connected by metal contacts.[1] Silicon chips form the random access memory in computers, whether desktop, laptop or tablet and in smartphones. Permanent magnets, for example, with compositions based on neodymium-boron-iron are present in computer hard drives. Powerful magnets are also a key component of scanners used for magnetic resonance imaging (MRI) in hospitals. Displays in smartphones and computers can contain light-emitting diodes (LED) and the screens contain a transparent electrically conducting indium tin oxide coating on the inner surface. Both smartphones and portable computers use light-weight rechargeable lithium-ion batteries for improving battery life. Lanthanide elements, for example, europium, terbium and praseodymium are used as phosphors to generate colours on the screens.[2] Consumer products such as televisions and in-vehicle navigation systems use liquid crystal displays (LCD) but as liquid crystals are passive and do not emit light, a backlight is used. The latter can make the screens bulky. Incorporation of light-emitting diode sidelights into television screens containing liquid crystals enables production of thinner screens, which is a more desirable feature for the consumer.

While metal alloys form the basis of light-emitting diodes, they have a completely different application in gas turbines. Oxidation- and corrosion-resistance of nickel-based superalloys enable these materials to be used at high temperature and pressure, in particular as turbine blades in gas turbines in aircraft and power generation plants. Toughness, ductility, low rate of creep and stability at high temperature are desirable properties of the superalloys. However, the gas temperature may reach 1500°C and, to prevent blades from melting, their surfaces are coated with a protective ceramic barrier. A common thermal barrier material is yttria-stabilised zirconium oxide deposited by thermal spraying. Turbine blades have complex structures incorporating internal cooling passages through which cool air is passed. Fabrication of complex components such as turbine blades is attracting attention from developments in manufacturing techniques such as 3D or three-dimensional printing.[3] Light-weight composite materials increase the performance of jet engines.[4] For example, silicon carbide fibres in titanium compressor discs and drums as well as polymer matrix composites in the outer casing (nacelles) of engines.

The worldwide market for flame retardants is large; currently it is approximately $4 billion and it is estimated to grow to $5.8 billion by 2018.[5] The prominent material for flame retardants in extruded and expanded polystyrene foam used in building insulation is hexabromocyclododecane. However, concerns over the toxicity of brominated flame retardants means that we are moving towards a global ban on this material. Flame retardants represent a fertile area for the development of economic and effective alternative materials as bromine-based materials are phased out. Examples of alternative materials are phosphorus-based retardants such as metal phosphinates, phosphonates and phosphate esters,[6] as well as inorganic-based materials such as alumina trihydrate (gibbsite) and clays including montmorillonite. Applications of flame retardants range from, for example, printed circuit boards, injection-moulded products, textile coatings, fire-proof paints and electronic components (*e.g.* cables and plugs). When phosphorus-based retardants are used in, for example, polyurethane foams and textiles they form a protective char layer that protects the product from attack by oxygen and heat. When incorporated into thermoplastic cables, as an example, inorganic flame retardants such as alumina trihydrate release water that cools and dilutes the flame zone.

Ionic liquids are salts that are liquid at room temperatures[7] and they are obtained by reducing the electrostatic forces between ions in a solid by incorporating bulky asymmetric cations such as pyridimium cations into the solid. Properties of these materials, such as melting point, viscosity and toxicity can be tailored by use of combination of different anions and cations. Potential applications of ionic liquids include: (i) solvents in synthesis as these liquids have low vapour pressures; (ii) formation of pharmaceuticals such as ampicillin as an ionic liquid avoiding polymorphs, that is, undesirable crystalline forms with limited pharmacological activity; (iii) nanoparticles of lanthanide-based phosphors as coatings in energy-saving light bulbs; and (iv) a commercial process for scrubbing mercury vapour from natural gas streams by absorption into an ionic liquid adsorbed on a solid support. Although ionic liquids have been known for over 100 years, increasing interest in them is represented by the filing of more than 1000 patents.[7]

Materials have a key role in renewable energy strategies. Hence, in solar or photovoltaic cells, the conventional material is crystalline silicon in the form of panels mounted in modules.[8] The latter need to be robust in order to survive exposure to extremes of weather and to operate reliably and efficiently for a number of years. An alternative approach to the use of crystalline silicon is the development of thin film photovoltaics in which a photosensitive material, for example, copper indium gallium diselenide is deposited as a thin layer onto a low-cost substrate such as glass, stainless steel or plastic. Another approach is use of dye-sensitised solar cells.[9] Here, a dye traditionally based on ruthenium is deposited as a coating onto a semiconductor material such as titanium dioxide. Unlike cells based on crystalline silicon, dye-sensitised cells are not sensitive to the angle of incoming light and do not operate only in bright sunlight conditions.

Potential disruptive technologies can rely on materials and their properties. The technique of 3D or three-dimensional printing is attracting much publicity in the public domain. This additive manufacturing technique allows for the manufacture of three-dimensional parts layer by layer. Applications of 3D printing include jewellery, dental implants, medical implants such as lower jaw implants, components for aircraft and even foodstuffs such as chocolate confectionery. The success of the technique depends in part on the availability of plastics that can be converted into molten droplets and ceramic and metal powders that can be 'printed' to form the component. Graphene is a two-dimensional sheet of carbon that is one atomic layer thick and is the thinnest and strongest known material. It was isolated in 2004 at the University of Manchester and, as with 3D printing, graphene has received much publicity and often referred to as a *'wonder material'*. Graphene represents a disruptive technology with a wide range of potential applications that exploit its mechanical, electrical and optical properties,[10] for example, hardness, strength, electrical conductivity and optical transparency. Examples of these potential applications include use in flexible electronic touch screens, electrodes in lithium-ion batteries and as a future replacement for silicon-based electronics. Graphene is a fertile area for patenting activity and 5174 patent publications have been issued (about 1% of them are associated with the United Kingdom).[11] A third potential disruptive technology is synthetic biology,[12] a multidisciplinary field with contributions from natural product chemistry, materials science, biotechnology and genetic engineering. Potential applications include medical diagnostics, tailored microbes to produce medicines, fuels such as butanol from sugars, polymeric structures for creating artificial life forms and regenerative medicine, that is, the replacement of diseased with healthy tissue. The global market for synthetic biology products is estimated to be £7 billion by 2016.[13]

A wide variety of materials are used in surgical implants; metal, polymer and ceramic. For example, in coated stents for drug delivery, hydroxyapatite coatings on metallic hip implants, polymers in inter-aortic balloons, metallic stents and spinal implants. Nitinol, a shape memory alloy formed of equal amounts of nickel and titanium is used in stents.[14] When inserted into diseased arteries it expands and holds open the artery. Smart electronic devices based on thin, flexible silicon circuitry and which can be attached to body organs and skin and transmit vital sign measurements (*e.g.* temperature, blood pressure or heart rate) are also under development.[15] Smart clothes or smart textiles that have a function other than the traditional role of textiles are attracting attention.[16,17] For example, a thin layer of titanium dioxide on the surface of textile fibres can photocatalytically oxidise stains or dirt attached to clothing to carbon dioxide and water.[16] Flexible plastic strips containing sensors can be interwoven with sensors to measure body temperature or to produce glowing light displays in clothing though embedded light-emitting diodes.[17]

1.3 PATENTS

1.3.1 Sources of Technical Information

Patents describe inventions, namely apparatus, processes (*i.e.* methods) or systems, in particular the technical features of inventions. Materials, processes for their preparation, their properties and uses constitute inventions that may be patentable. An inventor will be able to answer at least one of the following questions:

    (i) What does my invention do?
    (ii) How does it do it?
    (iii) What is my invention made of?
    (iv) How is it made?

National patent offices hold information on granted patents and published patent applications on databases. For example, the United States Patent and Trademark Office (USPTO) issues newly granted patents and newly published applications on a weekly basis.[18] The United Kingdom Intellectual Property Office (UKIPO) records very brief details for unpublished applications, about six weeks after filing, in *Patents and Designs Journals*.[19] International applications are issued weekly by the World Intellectual Property Organization (WIPO) and these applications are identified by a prefix WO, with their publication number.[20] International applications are sometimes erroneously referred to as world patents but there are in fact no world patents. Espacenet is a database operated by the European Patent Office and a key feature of this database is that patent documents can be obtained from many jurisdictions, for example, China and Russia.[21] Commercially available software is available that simplifies the searching process for these databases.

The patent literature is vast. For example, the first United States patent was granted on 31 July 1790 and there are now over eight million granted United States patents.[22]

1.3.2   The Structure of Patent Documents

All patent documents have the same structure:

- Front page. This contains bibliographic data that is useful for market intelligence. That is, the title, abstract, inventor names, applicant (owner) name, patent classification, priority date, filing date and publication date. Numbers in brackets on the front page, for example '[54]', are INID codes (Internationally agreed Numbers for the Identification of bibliographic Data). If the document is in, say, Russian or Japanese then these codes will enable a reader to identify the sequence of dates used in the filing process. INID codes are listed in reference 23.
- Background to the invention. This section gives a description of the prior art, that is, earlier published documents of potential relevance to the invention. These documents can be patent documents, journal articles or anything in the public domain. This section is usually very comprehensive for US applications and grants and is a good source of references when finding out about a new field of technology.
- Summary of the invention. This section gives a brief description of the invention. Note that one patent describes one invention.
- Embodiments. Several ways of producing the invention are described along with the preferred way.
- Drawings. Labelled drawings are supplied at the front or back of the document.
- Claims. The claims are the legal part of the patent document and they describe the scope of the rights given to the applicant (owner). There are independent claims, such as claim 1, that do not link to other claims, as well as dependent claims that link to other claims. Patent infringement can occur when someone has a product, method or system that is described in part by at least one claim of another patent. Note that claims described in one patent do not infringe claims described in another patent.

The description, drawings and claims for a patent application are the patent specification. Only features described in the specification can be claimed.

When a patent specification is prepared, the inventors frequently refer to or cite earlier published documents that they consider to be related to their invention as described in the text. These earlier publications that can be any document in the public domain, not just patent documents, are known as backward citations. The inventors aim to distinguish their invention from the technical features referred to in the cited documents. Additional citations are brought to the attention of the inventors either through a search report or during the examination process. Backward citations are listed in published patent documents, for example, on the front page of granted United States patents. In addition, a published patent document can be cited by documents published at a later time. In this case these citations are known as forward citations. The frequency of citation of patent document is a qualitative measure of its importance in a particular technical field. Thus, for a particular patent portfolio, the most cited documents may represent key documents in a technical field. It is not unknown for some documents to be cited five hundred times or more.

There is no restriction on the length of patent documents and they can be over 100 pages. The language used in patent documents can appear to be arcane and sentences can be long. Studying the documents can be a time-consuming process. Words and phrases in a specification do not have to have the same meaning they would have in everyday life. Their meaning is determined by the way they are defined in the specification. The phrase used by Humpty Dumpty in the book *Through the Looking Glass* by Lewis Carroll can seem relevant to the interpretation of patent documents, he says, 'When I use a word, it means just what I choose it to mean, neither more nor less'.

A quick overview of a patent document can be obtained by studying the abstract, the summary of the invention and independent claim 1.

1.3.3   The Filing Process

A brief description of the filing process is given here. A patent specification is prepared and when carried out in the United Kingdom it is then filed, that is, deposited, at the United Kingdom Intellectual Property Office (UKIPO). A filing date is issued by UKIPO. About six weeks after filing a very short description of the invention is listed in the *Patents and Designs Journal* by UKIPO. Additions to the specification can be made by the inventor in the twelve months following initial filing and the amended specification is then re-filed; added material is associated with the new filing date. The patent owner, that is, the applicant or assignee, can request a search report from UKIPO and this report will indicate prior art documents that may affect patentability. Twelve months after the first filing date, specifications that allow for added material are combined or cognated and the applicant has to decide what route to follow in order to obtain granted patents. As an example, an international patent application can be made that will be published eighteen months after the initial filing and which claims priority from all of the filing dates issued by UKIPO. International search and examination can be requested. National and regional filings are made thirty months after the priority date. Hence, a regional filing at the European Patent Office to cover countries in the European Patent Convention but for other countries such as China or Japan a translation is required. The European patent application is examined over a period of around four years, a grant fee is paid and the patent is then validated in selected countries. In other countries, for example, China, Japan and the USA, grant and renewal fees are paid to keep the granted patent in force. In the USA renewal fees are paid 3.5, 7.5 and 11.5 years from grant and the lifetime for a granted patent is twenty years from the filing date. Introduction of a single harmonised European patent system is anticipated in 2014.[24] This will limit the costs of obtaining patents in countries that have signed up to the system as patents in English, French or German will be valid in those countries, thus reducing the cost of translations.

Patents are a form of intellectual property originating from creations of the mind. Other forms of intellectual property include copyright, trade marks, designs, databases and trade secrets. The following criteria are critical for obtaining a granted patent. First, the invention has to be novel, that is, not disclosed to the public at the time of filing the patent application.[25] In addition, the invention has to be inventive or non-obvious, so that it could not be predicted how the invention works by consideration of published documents in the public domain before the filing date. Besides the criteria of novelty and non-obviousness, the invention has to be capable of industrial application. A patent conveys a negative right to its owner, that is, the owner has the right to prevent others from practicing the invention, including manufacturing or importing goods for sale into a country where the patent is enforced. The patent owner does not have the automatic right to exploit the invention.

The importance of seeking specialist advice from qualified practitioners on aspects of novelty and inventiveness, especially when patent cover is sought is stressed again. References 26 and 27 give an overview of intellectual property for innovators and research managers.

1.3.4 Patent Infringement

Patents have attained a higher profile among the general public over the last five years, mainly because of the patent infringement lawsuits in mobile phone technology. These lawsuits are frequently referred to in the popular press, on radio and television and on websites. Infringement relates to manufacturing in or importing an invention into a country where a patent protects the invention.[23] The technical complexity that can arise in infringement lawsuits is illustrated in a case involving Samsung Electronics (UK) and Apple Inc,[28] in the United Kingdom, relating to tablet computers, while reviews of infringement cases in the USA are given in references 29 and 30. Lawsuits are held before a jury in the USA but only before a judge in the United Kingdom. Non-practicing entities (NPEs) are individuals or firms who own patents but do not directly use their patented technology to produce goods or services.[29] Instead, NPEs assert their patents against companies that do produce goods and services. Non-practicing entities are sometimes referred to as 'patent trolls', although when used in this way the phrase has derogatory overtones. It has been estimated that, while business spending on research and development in the USA was $247 billion in 2009, the direct costs of patent assertions by non-practicing entities in 2011 was $29 billion. Direct costs include the cost of external legal services and license fees, but indirect costs not included in this figure include loss of market share, a diversion of resources and delays in new products launched by the defendant.

The scale of litigation by NPEs in the USA is indicated by the 5842 defence cases mounted by 2150 companies in 2011.

Patents are a very valuable form of property, as shown by strategic intellectual property acquisitions in 2011.[30] For example, the 'Rockstar Group', a consortium of buyers including Apple, Microsoft, Research in Motion and Sony, acquired a portfolio of 6000 patents from the now defunct Nortel Networks in July 2011. Shortly after this purchase, Google acquired Motorola Mobility, reportedly for its extensive portfolio of 17,000 patents to protect the Android operating system from patent lawsuits. The risks of patent infringement, as well as the rewards of enforcing patent rights are illustrated by damages awarded in initial adjudication although awards are sometimes reduced at a later date.[30] For example, Bruce N. Saffran was initially awarded $593 million in a lawsuit involving Johnson & Johnson on drug-eluting stents in 2011.

Light-emitting diodes are described in Chapter 2. This is an area of rapid technological change that is reflected in recent lawsuits in areas as diverse as metallization technology, modular lighting products and power control circuits for light-emitting diodes.[31]

1.4 PRACTICAL GUIDE TO PATENT SEARCHING

Patent searches are carried out for various reasons. A patent attorney will want to know if the claims of any published patent documents describe the technical features of his or her client's invention in order to assess the possibility of patent infringement. A pharmaceutical company might identify the patenting activity of a competitor in a technical area of interest to the company and this information may be relevant to a possible acquisition. An employer may want to know if a possible new employee is a prolific inventor as measured by the number of patent documents associated with that person. Anyone starting a company based on a patentable idea might want to know if there are competitors with granted patents or applications in the same field. A university researcher assembling a proposal for funding from a government organisation may want to check the patent literature to ensure the proposed work is novel. A doctoral student in chemistry or materials science who has carried out extensive searches of literature published in scientific journals might want to carry out a brief search of the patent literature in the area of interest. The United Kingdom Intellectual Property Office will carry out, for a fee, a search of potentially relevant prior art documents including patent documents when an application is filed, in order to give an idea of documents that are relevant to the novelty and inventiveness of the invention.

Patent searches can be carried out by using publicly accessible databases or by using licensed software. In both cases, the value of the results of searches is very dependent on the criteria used for the searches. Anyone who is new to patent searching may wonder how best to get started. Although this monograph is not an instruction manual on patent searching, some practical guidelines are given here.

One approach to patent searching is to use a keyword or phrase, for example, 'ionic liquid' or 'hydrogel'. In order to allow for plural versions of words it is useful to truncate the words and use 'hydrogel*' or 'ionic liquid*' where the asterisk allows for hydrogels or ionic liquids.

One can search different parts of the specification with keywords or phrases, that is, (i) the title, (ii) the abstract, (iii) the claims, (iv) title/abstract/claims or (v) the full specification. Searching the full specification will usually produce a far larger set of documents than searching on the other parts of the specification. If one wants a focused set of documents in an area of interest, then searching just on the title or title/abstract/claims will achieve this. If the area of interest is known to be a very active area for patenting, then a search on the titles will limit the size of the results set. Often, the number of results returned in a search is limited and this number is set by the provider of the software, either publicly accessible or licensed. There may be scope for searching over a time interval. Hence, you can search the keywords for documents that have been filed in the preceding twenty years. The twenty-year period is convenient, as the lifetime of patents is twenty years from the filing date. However, for a field such as graphene, which was isolated as recently as 2004, it is anticipated that the majority of patent documents will have been filed in the last ten years. It may be possible once a set of results is obtained from a search to reduce the document set to families. A patent describes one invention and a separate patent is required for each country where the patent is to be enforced. So, if an applicant has a patent for a specific invention in the United Kingdom, Japan, France, Spain and the USA, then these five documents are known as equivalents in one patent family.

There are drawbacks to keyword searches. Firstly, there may be spelling mistakes in the patent specification. This can be a problem when searching a surname, for example, meaning that relevant documents may not show up in searches. In addition, patent specifications can be constructed in ways that make it difficult to identify the key features of the invention. For example, graphene is sometimes referred to as an exfoliated graphite oxide. If a patent specification referred to exfoliated graphite oxide throughout, with no mention to graphene then a search on the keyword graphene will not identify the document. If a material has a common name and a chemical name, then keyword searches may not identify a complete set of documents unless all variations on the name are searched for. Hence a flame retardant can also be referred to as a fire retardant. Early documents on three-dimensional printing did not refer to this phrase and this technique had been described as stereolithography in the patent literature, so that a search on three-dimensional printing may not find documents on stereolithography.

Another way of carrying out a patent search that avoids the use of keywords is to do a classification search. When a patent specification is filed at a national patent office, that office assigns a patent classification to the document. The classification describes the subject area of the invention and is shown on the front page of the patent document. Although changes to classification systems take place periodically, it is important to stress that searches can be carried out based on patent classifications. In addition, searches can be carried out using a combination of classifications and keywords.

While licensed software simplifies the searching process, not everyone with interests in patents will have access to specialised software and they will have to analyse a set of documents manually. It can be useful to scan the titles and abstracts in a document set. Those documents that seem to be of relevance should be looked at in more detail. In the first instance, a study of the background section of patent documents, particularly those that were filed initially in the USA often give a good description of prior art documents. For example, in the field of ionic liquids, pioneering work by Wier and Hurley in the 1940s is often referenced in later documents. However, it is not obvious in, for example, the area of graphene what documents should be studied in detail. One way forward is to divide the documents into smaller sets that describe specific applications and then concentrate on possible representative documents. Documents that are highly cited may be relevant documents in a field. This approach will not be rigorous enough for a patent attorney who needs to identify key published documents relevant to the client's invention but it should be appropriate for anyone who wants an overview of patenting activity in their area of interest. In addition, if a researcher knows of key players in a field of interest then a search on the surnames will identify potentially relevant documents.

Patent searches can return documents that are not in English. This can be the case for international applications and European patent applications and grants. Equivalent documents in English can be identified, if available, from the Espacenet site of the European Patent Office.

It is appropriate at the end of this introductory chapter to highlight the pioneering work of Stephanie Kwolek who died in June 2014 at the age of 90. Although this name will be unknown to the majority of the general public and also among the scientific community, many people will have heard of lightweight bullet-proof vests made out of the synthetic polymer Kevlar®. Stephanie Kwolek is credited as the inventor of Kevlar® and is named as an inventor or co-inventor of seventeen granted United States patents. Kevlar® is a crystalline spun fibre derived from polyamide solutions that contain rod-shaped liquid crystalline phases. Properties of Kevlar® include low density, high stiffness, high tensile strength with melting points up to around 500 °C and these properties are attractive for applications including high-temperature electrical insulation, parachutes, cords for tyres and conveyor belts. Coatings and self-supporting films of Kevlar® can be produced as well as fibres. A detailed description of the preparation of Kevlar® fibres is given in the patent literature.[32–36]

1.5   SUMMARY

The patent literature is vast, for example, there are over eight million granted United States patents. Physics, chemistry, materials science, engineering and mathematics underpin features of inventions described in the patent literature. Materials including metals, ceramics and polymers feature prominently in many technology areas and patent literature can be used to monitor developments in these technologies. The public profile of patents has increased over the last five years, due in part to patent infringement lawsuits relating to mobile phone technology.

REFERENCES

1. J.-H. Lee, J.-H. Cho, J.-S. Choi and D.-J. Lee, Spin-on glass composition and method of forming silicon oxide layer in semiconductor manufacturing process using the same, *United States Patent*, 7 270 886, 2007.
2. R. Lawley, What's in your laptop?, *Mater. World*, 2013, **21**(6), 31.
3. T. Probert, Blade runner: improving gas turbine coatings, *Mater. World*, 2013, **21**(11), 38–41.
4. M. Felice, Materials for aeroplane engines, *Mater. World*, 2013, **21**(5), 52–53.
5. N. Moran, Phasing out fire retardants, *Chem. World*, 2013, **10**(8), 50–53.
6. M. Burke, A new market for retardants, *Chem. Ind.*, 2010, **74**(10), 17–19.
7. L. Reade, Charged with success, *Chem. Ind.*, 2013, 77(10), 42–45.
8. N. Eisberg, Shining a light on reliable PV, *Chem. Ind.*, 2012, **76**(10), 24–27.
9. P. Broadwith, Dyeing for a place in the sun, *Chem. World*, 2012, **9**(6), 52–55.
10. L. Asfa-Wossen, Graphene: critical mass, *Mater. World*, 2013, **21**(12), 12–13.
11. R. Lawler, Nanotechnology under the microscope, *Mater. World*, 2013, **21**(4), 24–25.
12. M. Peplow, Bursting with life, *Chem. World*, 2013, **10**(9), 39.
13. Technology Strategy Board, *A synthetic biology roadmap for the UK*, 2013, (http://bit.ly/13Yi7cS).

14. F. Case, Magical mixtures of metals, *Chem. World*, 2013, **10**(11), 54–57.
15. A. King, Clinically smart sensors, *Chem. Ind.*, 2014, **78**(2), 32–35.
16. N. Notman, Clothing gets smart, *Chem. World*, 2012, **9**(10), 58–61.
17. M. Burke, Smart clothes, *Chem. Ind.*, 2013, **77**(4), 20–23.
18. United States Patent and Trademark Office. http://www.uspto.gov.
19. United Kingdom Intellectual Property Office. http://www.ipo.gov.uk.
20. World Intellectual Property Organization. http://www.wipo.int/portal/index.html.en.
21. Espacenet. http://worldwide.espacenet.com.
22. S. Hopkins, Improvement in the making of Pot Ash by a new apparatus and process, *United States Patent*, X000001, 1790.
23. S. van Dulken, *Introduction to patent information*, British Library, London, 3rd edition, 1998.
24. N. Moran, Single EU patent agreed, *Chem. World*, 2013, **10**(2), 14.
25. D. Segal and P. Tolfts, Keeping your IP fit and healthy, *Eur. Med. Device Technol.*, 2013, **4**(1), 44–48.
26. J. McManus, *Intellectual property: from creation to commercialisation*, Oak Tree Press, 2012.
27. H. Jackson Knight, *Patent strategy for researchers and research managers*, Wiley, 2012.
28. His Honour Judge Birss QC, Samsung Electronics (UK) Limited v Apple Inc. Case number HC 11 C03050, Neutral citation number [2012] EWHC 1882 (Pat). In the High Court of Justice, Chancery Division, Patents Court, 18th, 19th June 2012.
29. J. Bessen and M. J. Leurer, *The direct costs from NPE disputes*, Boston University School of Law Working Party number 12–34 (25 June), revised 28 June 2012.
30. L. Ansell, R. Arad, M. Arnold, C. Barry, A. Johnston and A. Parent, *2012 Patent litigation study*, Pricewaterhouse-Coopers LLP, reference NY-13-0033, 2012.
31. J. Maguire and D. Segal, IP strategy plays key role in LED business development, *LEDs Magazine*, 2012, 57–60.

32. H. W. Hill Jr., S. L. Kwolek and P. W. Morgan, Polyamides from reaction of aromatic diacid halide dissolved in cyclic non-aromatic oxygenated organic solvent and an aromatic diamine, *United States Patent*, 3 006 899, 1961.
33. H. W. Hill, S. L. Kwolek and W. Sweeny, Halogenated aromatic polyamides, *United States Patent*, 3 349 062, 1967.
34. H. W. Hill Jr., S. L. Kwolek and W. Sweeney, Aromatic polyamides, *United States Patent*, 3 380 969, 1968.
35. S. L. Kwolek, Wholly aromatic carbocyclic polycarbonamide fiber having orientation angle of less than about 45°, *United States Patent*, 3 819 587, 1974.
36. S. L. Kwolek, Optically anisotropic aromatic polyamide dopes, *United States Patent*, 30 352, 1980.

CHAPTER 2

Light-emitting Diodes

2.1 INTRODUCTION

Spectators at the opening and closing ceremonies of the 2012 Olympic Games, held in London, as well as a worldwide television audience, witnessed spectacular light displays. More than 70,000 LED (light-emitting diodes) modules (or pixel tablets), each containing nine LEDs, were positioned next to each seat in the Olympic Stadium.[1] Patterns, text and animated effects in the crowd section were generated using these modules. Some of the television audience will have received the broadcast from a fibre optic network. There is a common theme between the light displays at the opening and closing ceremonies and transmission of the signal along optical fibres, namely semiconductor materials. The latter are crucial to the operation of light-emitting diodes and are also an important element in communication lasers, which are used in the transmission of signals in fibre optic networks. The spectators may not have appreciated that the emergence of light-emitting diodes onto a worldwide stage represents the culmination of research, development and manufacturing efforts over more than a hundred years that began with observation of a 'curious phenomenon'. In this chapter, a short account of semiconductor materials is given, followed by a journey back to the beginning of the 20th century and the

Exploring Materials through Patent Information
By David Segal
© David Segal, 2015
Published by the Royal Society of Chemistry, www.rsc.org

discovery of electroluminescence. This is followed by an account of the development of experimental methods for the preparation of semiconductors for use as light-emitting diodes and in communication lasers. The study of semiconductors and lasers is a multidisciplinary area and detailed accounts of these areas are given in the patent literature. Interested readers are recommended to consider the patent literature as a relevant source of information in these areas.

2.2 SEMICONDUCTOR MATERIALS

Reference to semiconductor materials is made throughout this chapter and a very brief outline is given here. Semiconductors have an electrical resistance between conductors and insulators. In its pure form, silicon is intrinsically an insulator. If a Group V element such as phosphorus is introduced into silicon, an excess of free electrons is introduced into the silicon lattice. Phosphorus is known as an n-type donor. If a Group III element such as boron is introduced into silicon, an excess of holes is introduced into the silicon lattice. Boron is a p-type acceptor. The doping process allows electrons to migrate from the valence band to the conduction band of the material, increasing the electrical conductivity. This transition is shown qualitatively in Figure 2.1. Under an electrical potential interaction of electrons and holes at the interface of p-type and n-type regions, electroluminescence results, that is, emission of light whose wavelength depends on the value of the bandgap energy.

2.3 LIGHT-EMITTING DIODES

2.3.1 The Early Years

In 1907, Henry Round published a short note in *Electrical World*, in which he referred to a curious phenomenon.[2] In his words: 'on applying a potential of 10 volts between two points on a crystal of carborundum, the crystal gave out a yellowish light'. This bright glow from a carborundum (silicon carbide) diode was the first observation of electroluminescence. In order to give Round's observation an historical perspective, his publication came out just two years after the announcement of Einstein's Special Theory of Relativity (1905), seven years after publication

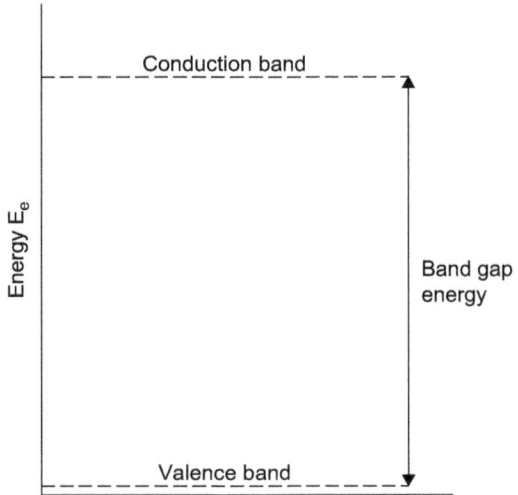

Figure 2.1 Representation of valence and conduction bands in a semiconductor.

of Planck's quantum theory (1900) and some years before the advent of quantum mechanics, as described by Schrodinger's wave equation (1926) and Heisenberg's uncertainty principle (1927). These four concepts have had a central role since their announcement in describing the structure and properties of solid-state materials, including semiconductors such as silicon carbide. However, in 1907, processes occurring in solid-state materials responsible for light emission and semiconductor band theory were not well understood, so that Round's use of the phrase 'curious phenomenon' is appropriate for his time.

Reference should be made to Oleg Vladimirovich Losev,[3–5] a Russian radio expert and inventor who published widely between 1927 and 1941 in Russian, British and German journals. For example, in 1927 he observed light emissions from zinc oxide and silicon carbide crystal rectifier diodes used in radio receivers when a current passed through them. In addition, he determined the current threshold for the onset of light emission from the point of contact between a metal wire and a silicon carbide crystal, recording the spectrum of this light. Although Losev's work did not receive much attention up until his death at the age of 39 during the blockade of Leningrad in 1942, he can correctly be credited as an inventor of the light-emitting diode.

2.3.2 Advances in Fabrication

Further developments in LEDs were delayed until methods for semiconductor fabrication with high chemical purity had been optimized. Rosi and Braunstein produced crystalline semiconductor alloys of gallium arsenide (GaAs) and indium arsenide (InAs) with GaAs contents between 6–50 mole %.[6] These alloys had a sharp transmission edge for infra-red wavelengths. Charges of gallium and indium were contained in a quartz boat at one end of a quartz tube; a charge of arsenic was placed at the other end of the tube furnace. Typical furnace temperatures were 950–1250 °C for the gallium indium alloy and 550–650 °C to vaporise the arsenic. Under these reaction conditions arsenic vapour diffuses into the Ga/In melt. An alternative preparation involved forming a melt between GaAs and InAs.

In later work Braunstein and Luebner produced infra-red emission from a solid-state device under an applied electric field.[7] A p-n junction was formed from a solid dot of lead-antimony on a p-type single crystal of germanium. This circuit is equivalent to a Ge diode with a p-type wafer, n-type dot and p-type wafer surface. Germanium could be replaced by gallium arsenide, indium phosphide, gallium phosphide, gallium antimonide or silicon carbide. Radiative emission in these solid-state devices is by electron-hole recombination.

A note on terminology is useful for clarification.[8,9] The electrical resistivity of elemental and compound semiconductors may be changed by the introduction of impurities known as dopants or activators, which are introduced by methods including ion implantation, diffusion of dopants or by epitaxial growth of a crystalline semiconductor layer containing one dopant on top of a crystalline layer containing a different dopant. N-type dopants, referred to as 'donors', are typically Group V elements such as arsenic, phosphorus and antimony and introduce an excess of free electrons into the semiconductor. However, p-type 'acceptors', typically Group III elements such as In, Al, Ga and B introduce an excess of holes or electron deficit into the semiconductor. A p-n junction is formed at the boundary between p-type and n-type semiconductors.

Both p-type and n-type semiconductors can be relatively conducting, although the p-n junction can be relatively non-conducting depending on the voltage between the semi-conductor regions. This non-conducting boundary layer contains fixed non-mobile ions and can be manipulated by forward or reverse bias where the term bias refers to the application of a voltage to the p-n junction. In forward bias, p-type semi-conductor is connected with the positive terminal and n-type semiconductor is connected with the negative terminal of the voltage source. Overall, electrons flow through the n-type region towards the junction and holes flow through the p-type region in the opposite direction towards the junction. Electron-hole re-combination takes place in the neighbourhood of the p-n junc-tion, causing electroluminescence. A localised electron-hole pair is known as an exciton.

In reverse bias, p-type semiconductor is connected with negative terminal and n-type semiconductor is connected with the positive terminal of the voltage source. Holes and electrons move away from the p-n junction and the depletion region widens. This movement increases until the potential across the junction equals the applied voltage and the current stops. The p-n junction acts as an insulator in reverse bias.

In summary, manipulation of the non-conducting boundary layer by an applied voltage allows the p-n junction to function as a diode under forward bias where electrons move in one dir-ection only. Electron-hole recombination in the vicinity of the p-n interface results in light emission so that semiconducting diodes form the basis of solid-state lighting technology. Re-combination and emission are associated with electrons occu-pying lower energy levels: the valence band. The colour (*i.e.* wavelength) of the light is determined by the bandgap energy of the semiconductor.

Biard and Pittman developed an infra-red emitting gallium arsenide diode under a forward biased voltage.[10] Of particular importance, the LED could be manufactured by existing tech-niques used for semiconductor fabrication. A wafer, about 1–2 cm^2 in area and approximately 500 μm thick was cut from a gallium arsenide crystal grown from a melt. N-type donors in-cluded tin, tellurium, sulphur and germanium, while zinc, used for the p-type layer was deposited by diffusion of zinc atoms from the vapour phase into the crystal. Lapping and aqueous etching of the wafer were carried out, after which it was diced into several gallium arsenide diodes. Electrical contacts are then deposited by, for example, evaporation and alloying techniques. Typical thicknesses of the n-type and p-type regions were 63 μm and 12 μm, respectively and application of a forward current bias by direct, pulsed or alternating current to the p-n junction produced infra-red emission by electroluminescence.

Advances in fabrication were made by Holonyak who developed methods for epitaxial deposition of compound semiconductors suitable for construction of electrical devices.[9,11] A substrate or seed such as gallium arsenide was placed at one end of an evacuated quartz tube. A source, gallium arsenide, gallium phosphide or indium phosphide was placed near the other end of the tube. A quantity of arsenic (up to 20 mg) or phosphorus (up to 50 mg) and a volatile metallic halide ($ZnCl_2$, $CdCl_2$, $CuCl_2$, $SnCl_2$, $MgCl_2$, $HgCl_2$ or $AlCl_3$) were positioned between the substrates and the source. Careful control of the temperature along the tube furnace in the range 600–1100 °C deposited epitaxial layers of GaAs, GaP and $GaAs_xP_{1-x}$ onto GaAs and GaP substrates, for example, $GaAs_{0.24}P_{0.76}$ that emitted red light in the visible spectrum.

Groves *et al.* prepared electroluminescent materials and devices based on nitrogen-doped epitaxial layers of $GaAs_{1-x}P_x$ deposited on single crystal GaP.[12] A stream of HCl in H_2 was passed over elemental Ga at 770 °C to produce a reactant vapour of gallium chloride. A second H_2 stream was mixed with AsH_3 and PH_3 and diethyl telluride. Both streams were combined and heated at 925 °C in a reaction zone. Vapours were then transported over a GaP substrate, when epitaxial deposition of $GaAs_{1-x}P_x$ took place at 825 °C. For example, the composition $GaAs_{0.235}P_{0.765}$ was grown to a layer thickness of 330 μm. Nitrogen was introduced into the layer by exposure of the latter to a mixture of NH_3/H_2 and the nitrogen-doped layer was about 18 μm in thickness. The N_2-doped layer was then exposed to Zn and P vapours to produce a p-type region and p-n junction. This diode emitted orange light (604 nm) and nitrogen-doped diodes had higher brightnesses and efficiencies than N_2-free alloys.

When GaAs was used as the substrate, the p-n junction was formed using $ZnAs_2$ as the diffusant. Use of combined gaseous streams produced an epitaxial layer of $GaAs_{0.525}P_{0.475}$ with a thickness around 192 μm. A nitrogen-doped layer about 12 μm thick was obtained by exposing this epitaxial layer to a NH_3/H_2 mixture. Diodes were made by diffusion of $ZnAs_2$ to 800 °C to form a p-region and a p-n junction and this composition was a red emitter (665 nm). Nitrogen-doped $GaAs_{1-x}P_x$ compositions had improved brightnesses at lower current densities than N_2-free alloys and emitted a range of colours, red→yellow→ green (650 nm–560 nm). For nitrogen-doped $GaAs_{1-x}P_x$ alloy compositions, red light emitting LEDs corresponded to the range $0.4 < x < 0.6$, while for yellow LEDs x is between 0.6 and 0.9. A typical cross-section of a semiconductor device fabricated from $GaAs_{1-x}P_x$ is shown in Figure 2.2.

The availability in the early 1970s of LEDs with emissions across the visible spectrum found a demand for numeric display arrays for pocket calculators and instrumentation. Light-emitting diodes replaced incandescent and neon lamps as status indicators in vehicles and household appliances to indicate a faulty condition.[13,14]

Uses for LEDs, particularly in lighting technology, were limited until bright blue-emitting semiconductors became widely available. Pankove demonstrated electroluminescence in an epitaxially deposited layer of GaN on an optically transparent sapphire substrate by passing d.c. current at ambient temperature between point contacts attached to the layer.[15] Vapour phase epitaxial techniques such as MOCVD (metalorganic chemical

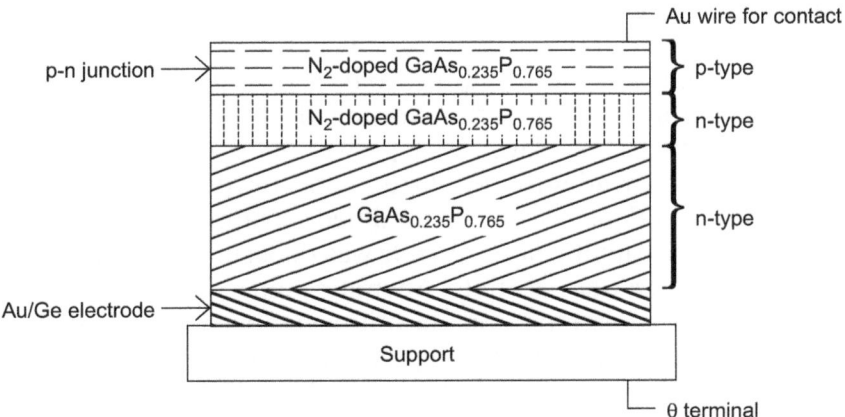

Figure 2.2 Schematic diagram of LED wafer.[12]

vapour deposition) were used to grow the layer. Acceptor impurities such as zinc were included in the GaN layer to form a p-type region. High concentrations of zinc produced blue light (470 nm), whereas lower concentrations produced a green light.

It was the use of magnesium as p-type dopant that produced the breakthrough for practical blue-emitting LEDs. Nakamura *et al.*[16] deposited a GaN buffer layer about 20 nm thick onto a sapphire substrate by MOCVD. Here, a mixture of trimethylgallium (TMG) and NH_3 in a hydrogen carrier gas was passed over the substrate at 510 °C. The substrate was then held at 1030 °C and Mg-doped GaN grown on top of the buffer layer from a mixture of TMG and cyclopentadienyl magnesium gas to a layer thickness of 4 µm, after which an annealing process was carried out at 800 °C at atmospheric pressure. Mg-doped GaN layers had good p-type characteristics, that is, a low electrical resistivity of 2 ohm.cm and hole carrier density of 2×10^7 cm^{-3}. In a variation of the experimental technique, a capping layer of GaN was deposited onto the Mg-doped GaN layer and then removed by etching after the annealing process. Use of the capping layer produced enhanced brightness for photoluminescence at 450 nm. In a further modification, Si-doped n-type $Ga_{0.9}Al_{0.1}N$ was grown was grown as a layer to a thickness of 0.8 µm on Mg-doped GaN using SiH_4 and trimethylaluminium as reactants. Electrodes were attached to each layer, after which the sapphire substrate was diced into pieces that were moulded into LEDs. Blue light emission with a peak wavelength at 430 nm and high brightness were obtained.

Electron-beam irradiation at 300 °C lowered the resistivity of p-type regions and increased the hole carrier density, which are desirable properties for light-emitting diodes. For example, a nitrogen-doped ZnSe layer was grown on a GaAs substrate. Electron-beam irradiation produced good p-type characteristics in the nitrogen-doped layer. In a modified experimental technique, a 0.1 µm thick capping layer of ZnSe was deposited onto nitrogen-doped ZnSe on a Gas substrate. After annealing, the capping layer was removed by etching and the nitrogen-doped layer had very good characteristics for blue light-emitting diodes, namely, resistivity of 0.6 ohm.cm and hole carrier density of 3×10^{18} cm^{-3}.

Blue emitting semiconductor compositions developed by Nagahama *et al.*[17] and Nakamura *et al.*[18] were $In_xGa_{1-x}N$

Table 2.1 The effect of composition on colour and wavelength of light-emitting diodes.[19]

Wavelength nm	Colour	LED composition
850	Infrared	GaAlAs/GaAs – Gallium aluminium arsenide/gallium arsenide
633	Super red	InGaAlP – Indium gallium aluminium phosphide
605	Orange	GaAsP/GaP – Gallium arsenic phosphide/gallium phosphide
585	Yellow	GaAsP/GaP – Gallium arsenic phosphide/gallium phosphide
6500 K	Pale white	SiC/GaN – Silicon carbide/gallium nitride
565	High efficiency green	GaP/GaP – Gallium phosphide/gallium phosphide
470	Super blue	SiC/GaN – Silicon carbide/gallium nitride

$(0 < x < 1)$ and $Al_yGa_{1-y}N$ $(0 < y < 1)$, where the wavelength of blue and green light emitted from active layers depends on the $In_x : In_xGa_{1-x}N$ ratio.

Table 2.1 shows the relation between semiconductor compositions and the emitted colour and wavelength.[19]

2.3.3 Packaging and Uses

Light-emitting diodes are fabricated in wafer form by methods used for manufacture of silicon-based integrated circuits. The wafer is reduced to chips by dicing and chip sizes are typically less than 1 mm^2 in area. The semiconductor chip is packaged for protection. The chip or die sits in a die cup, sometimes referred to as an anvil, that acts as a support and light reflector. This assembly is encased in epoxy resin in the shape of a transparent dome. While the diode emits light by electroluminescence, the shape of the dome determines the way radiated light is dispersed. LEDs generate heat under forward bias and, although this is much less than the heat generated from an incandescent bulb, a heat sink may be required for dissipation. Light-emitting diodes have no moving parts or filaments to break and have potentially longer lifetimes than conventional lighting, such as incandescent lamps.

White light can be made in several ways. Red, green and blue emissions from LEDs can be mixed to form white light.

Alternatively, ultraviolet emission from a light-emitting diode causes fluorescence in a white-emitting phosphor, a process that is similar to how a fluorescent tube generates white light. In the third method, a blue-emitting GaN-based LED (*e.g.* 450 nm) is covered by a layer of powdered cerium-doped yttrium aluminium garnet crystals. The LED chip emits blue light, part of which is converted to yellow light by the garnet phosphor. As yellow is a combination of red and green wavelengths, the mix of yellow and blue gives the appearance of white light. LEDs are not mono-chromatic light sources and emit with a small spread of wave-lengths around a peak emission and this results in a range of 'white' colour when red, blue and green wavelengths are mixed.

Light-emitting diodes represent an area of technology that is attracting considerable attention in the patent literature. For example, 3800 patent families on GaN and InGaN have been filed between 1992 and 2012, and this is an area of increasing activity as measured by the rate of filing.[20] Low energy consumption, long lifetime, robust construction, small size, fast switching for data transmission along with durability and reliability are at-tractive properties for applications. For example, automotive lighting, traffic signals, displays on digital clocks, interior lighting in aircraft, instrumentation, indicator lamps in elec-tronic equipment for on/off status, street, retail and home lighting as well as solid-state lighting generally. Many consumer products such as mobile phones, televisions, notebook com-puters, digital cameras and projectors have liquid crystal display (LCD) panels. LCDs are passive devices, not light-emitting de-vices, and they require a backlight unit for illumination. Tradi-tionally, this has been achieved by the use of a cold cathode fluorescent lamp (CCFL) that is relatively bulky and there is considerable interest in substitution of the fluorescent lamp by a LED panel. Light from the LED panel can illuminate the LCD unit from behind or from a sidelight. Examples of LED back-lights are described by Hamada and Yoshida *et al.*[21,22]

Efforts to produce an improved 'warm white' light continue. For example, Krames *et al.* first obtained blue light with wave-length between 450–470 nm from an LED with composition $BaM^I_{3-x-z}M^{II}_xSi_{6-a}Al_aO_{1-x+a}N_{10+x-a}:Eu_z$ with $M^I = Ba, Ca, Sr, Mg$; $M^{II} = La, Gd, Lu, Y, Sc, Ce, Pr, Sm; 0 < x < 1, 0 < z < 0.1, 0 < a < 3$.[23] Some of the blue emission was converted to red light by using a phosphor, for example, $Ba_2Ca_2Si_6ON_{10}$: Eu (1%) and another part of the blue emission was converted to green light using a phosphor such as $BaSrSiO_4$: Eu and the thiogallate $SrGa_2S_4$: Eu. Combination of red, green and blue gave white light.

Figure 2.3 shows schematically an LED chip assembly, where a representative LED composition is $In_{0.05}Ga_{0.95}N$ and the phosphor and the phosphor coating has the composition $Y_3(Al_{0.6}Ga_{0.4})_5O_{12}$: Ce, a green light emitter.[24]

One sign of an expanding technology is the degree of licensing that takes place. In the case of solid-state lighting over 200 licenses have been signed with Philips for its patent portfolio related to LEDs, according to Whitaker.[25]

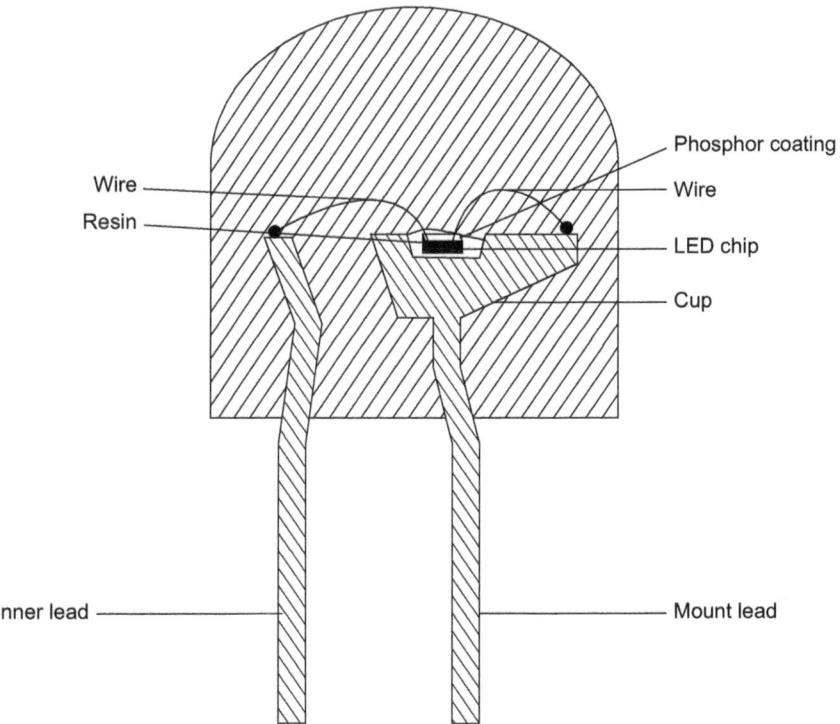

Figure 2.3 Schematic diagram of location of phosphor coating on a LED chip assembly.[24]

2.4 COMMUNICATION LASERS

2.4.1 Fibre Optic Networks

In recent years, broadband communications and public tele-communications networks using optical fibres have become widely available and are increasingly used.[26] In optical communication, information is encoded on light so that the propagated light waves convey the information between two sites, which can be across a room, or separated by thousands of miles. A fibre optic communication system can consist of terminals or users linked together *via* optical paths in a fibre optic network. Each terminal can contain a transmitter and receiver that form the basis of an optical transceiver, and the terminal receives or transmits light that is encoded with information. Hence, devices on an optical network can communicate with one another by sending and receiving optical signals. Semiconductor lasers are used as the light source in optical fibre communication systems and also as light sources in other optoelectronic systems including laser printers, optical measuring and compact discs such as DVDs (digital versatile disc).[27] The increasing use of the Internet and availability of services such as social media and video streaming has increased demand for the transmission of large amounts of information at low cost and high speed. The capacity of networks has been increased by the use of wavelength division multiplexing (WDM), in which many wavelengths and hence frequencies are transmitted simultaneously by optical fibres, and a variety of wavelengths can be obtained in tunable lasers in order to increase data capacity.[28] Short wavelength semiconductor lasers in the blue or blue-green regions are useful for optical data storage systems as the information density is inversely proportional to the square of the optical wavelength.[29]

2.4.2 Optical Fibres

Light is guided by total internal reflection in the core of an optical fibre.[30] The core is surrounded by a transparent cladding of lower refractive index. While light can be guided with near perfect efficiency, fibres can be affected in practice by losses. For example, losses can occur by (i) absorption arising from impurities in the core or cladding, (ii) Rayleigh scattering caused by

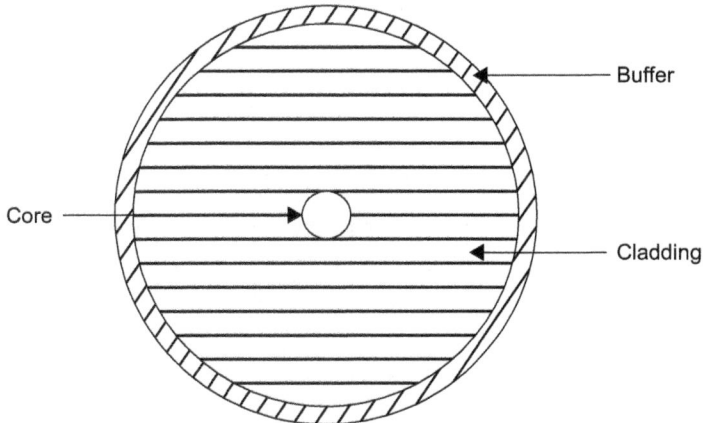

Figure 2.4 Schematic diagram for the cross-section of an optical fibre (not to scale).[30]

local variations in the refractive index of the core and cladding and (iii) imperfections in the core/cladding interface. Optical fibres are typically pumped by a laser from the end of the fibre, but have also been pumped from the side. A schematic diagram of an optical fibre is shown in Figure 2.4.

The optical fibre includes a silica-based core, a silica-based cladding and an outer buffer. The core and cladding are drawn from a preform that includes dopants, to provide a higher refractive index in the core relative to the cladding, which traps light inside the core. The buffer is a protective layer and is usually a plastic coating.

2.4.3 Quantum Wells

Quantum wells are formed from joining particular semiconductor materials having a thickness of around 10 nm or less, where effects of quantum confinement are important.[31] Under quantum confinement, the effective bandgap of the quantum well material increases with decreasing thickness. Semiconductor materials that act as cladding and barriers surround the wells. The heterojunction semiconductor layers form an active region or core into which carriers, both electrons and holes are injected and when a threshold current is reached, lasing occurs.

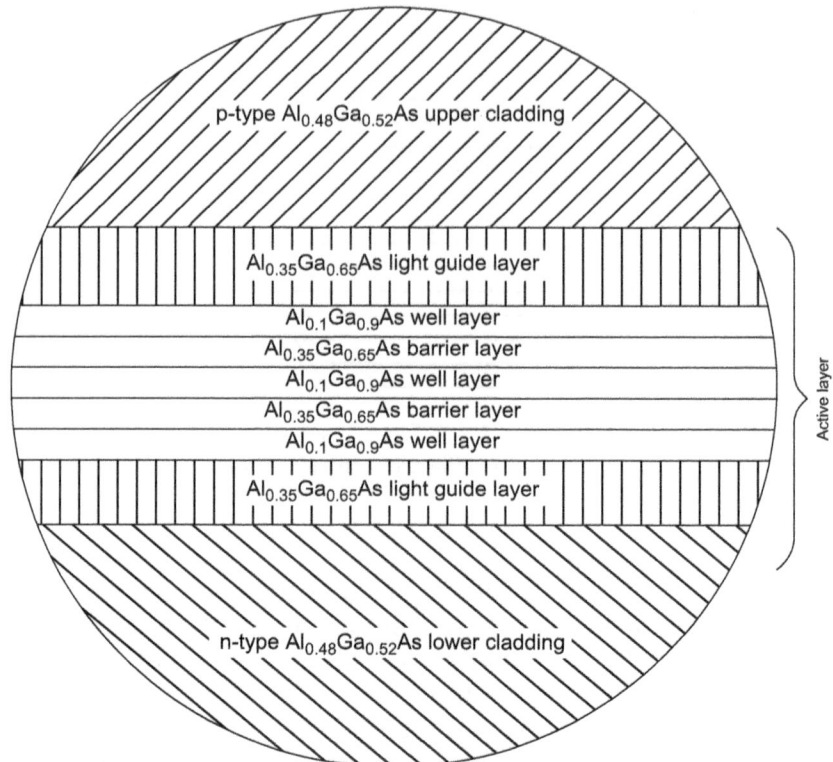

Figure 2.5 Schematic diagram for a quantum well structure.[32]

The frequency and wavelength of emitted light is a function of the structure and composition of materials in the active region. Quantum wells are key components in laser diodes as they strengthen electrooptical interactions by confining carriers to small regions. An example of a quantum well structure is shown in Figure 2.5.[32]

The quantum well active layer shown in Figure 2.5 consists of $Al_{0.35}Ga_{0.65}As$ light guide layers, $Al_{0.1}Ga_{0.9}As$ well layers and $Al_{0.35}Ga_{0.65}As$ barrier layers. Hence, three well layers are placed between two light guide layers with two barrier layers inserted between the three well layers. The thicknesses of the well layers, light guide layers and barrier layers are 8 nm, 20 nm and 8 nm, respectively. A lower cladding layer has a thickness 2–3 μm and consists of n-type $Al_{0.48}Ga_{0.52}As$ composition while an upper cladding layer has a p-type $Al_{0.48}Ga_{0.52}As$ composition.

A semiconductor device based on this well structure emits light with a wavelength of 0.98 μm.

Another example of a quantum well structure is shown in Figure 2.6 and this structure is applicable for a distributed feedback semiconductor laser diode.[33]

An n-type InP cladding layer with a thickness of 0.3 μm and carrier concentration of 1×10^{18} cm^{-3} is deposited on a n-type InP substrate and a diffraction grating layer was deposited on top of the cladding layer. The substrate has a thickness of 100 μm and a carrier concentration of 5×10^{18} cm^{-3} while the diffusion grating layer is formed of n-type InGaAsP and has a carrier concentration of 1×10^{18} cm^{-3} and a thickness of 0.05 μm. An n-InP layer with a carrier concentration of 1×10^{18} cm^{-3} and a thickness of 0.1 μm is deposited onto the grating to cover up a series of slits. An optical confinement layer has a composition n-AlGaInAs, a thickness of 0.1 um and a carrier concentration of 1×10^{18} cm^{-3}. The active layer, which has a quantum well structure of undoped AlGaInAs, has a thickness of 0.1 μm. Another optical confinement layer with a thickness of 0.1 μm, a composition of p-AlGaInAs and a carrier concentration of 1×10^{18} cm^{-3} is deposited on top of the active layer. A p-AlGaInAs

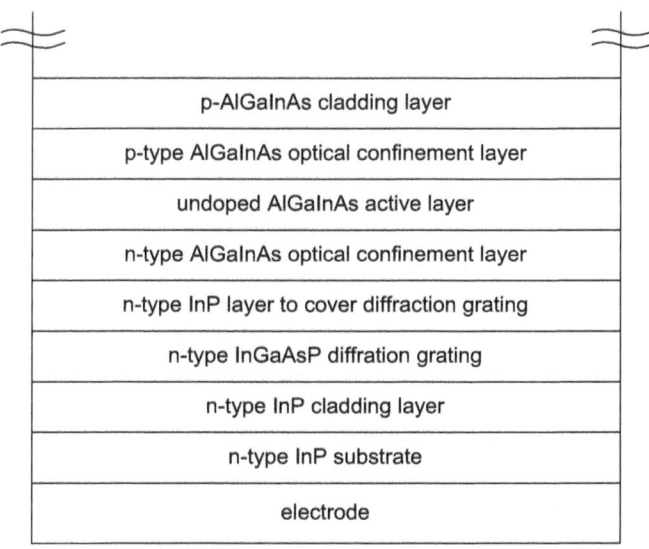

Figure 2.6 Schematic diagram showing a quantum well structure.[33]

cladding layer with a carrier concentration of 0.5×10^{18} cm^{-3} and a thickness of 0.1 μm is deposited onto the optical confinement layer. Deposition techniques for producing the quantum well are standard methods used in the semiconductor industry, for example, molecular beam epitaxy or metalorganic chemical vapour deposition. This example illustrates the intricate detail in quantum well structures.

In general, the active region forms a high optical gain region where lasing takes place. The cladding layers above and below the active region have a refractive index lower than the core refractive index and the cladding forms a waveguide that contains the optical modes in the core.[34] The guided optical modes propagate along the active region creating the laser beam. Further examples of quantum well structures in semiconductor lasers are given in references 35–41.

2.4.4   Tunable Lasers

Wavelength division multiplexing allows the capacity of a fibre optic network to be increased by transmitting encoded light waves of different wavelengths along the same optical fibre by means of tunable semiconductor lasers. Ideally, the range of wavelengths is between 1530 nm and 1610 nm.[42] There are three types of conventional tunable lasers: (i) systems with movable mechanical components such as diffraction gratings, etalons and prisms; (ii) systems for tuning the wavelength by adjustment of the temperature; and (iii) systems that use non-moving devices in the optical cavity for wavelength adjustment, such as magneto-optic devices, acousto-optic devices or current injection devices for selecting the wavelength. Methods that have been used for tuning the lasers for selected wavelengths include an acousto-optic tunable filter in the laser resonant cavity as well as the assembly of a gain region, a tuning region and an output coupler together on a substrate such as GaAs or InP.[42,43] Two types of optical resonators have been used in tunable lasers.[44] In a Fabry-Perot cavity, resonance takes place for specific optical wavelengths of light reflected between parallel reflective planes when the separation between the planes is an even multiple of the light wavelength. Resonance in ring resonators takes place for critical distances defined by the circumference of circular waveguides rather than the separation between reflective planes.

Use of planar fabrication methods has been used for ring resonators with no moving parts for use in tunable lasers, in particular a semi-ring Fabry-Perot resonator on a semiconductor substrate.[44] Another approach for fabrication of a tunable laser was to produce all components including the laser, optical amplifier and resonator on an epitaxial structure on a semiconductor substrate.[45] A diode laser pump source, for example, based on a neodymium-doped yttrium aluminium garnet laser pump converts electrical energy into a laser beam that can be switched between different optical paths and different laser cavities to produce multiple wavelength outputs from the cavities.[46]

2.5 SUMMARY

The study of light-emitting diodes dates back over a hundred years to the discovery of electroluminescence in 1907. Light-emitting diodes are semiconductors that convert electrical energy from the motion of electrons under forward bias into light by the process of electroluminescence involving electron-hole recombination at a p-n junction. The wavelength of emitted light is monochromatic depending on the bandgap of the semiconductor. A related application of semiconductors is in communication lasers for the transmission of encoded light waves down fibre-optic networks for delivery of broadband services to businesses and householders.

REFERENCES

1. T. Whitaker, LED lighting plays prominent role in Olympic Games, *LEDs Magazine*, 2012, http://ledsmagazine.com/news/9/8/9.
2. H. Round, *Electr. World*, 1907, **49**, 308.
3. O. V. Losev, *Philosophical Magazine*, 1928, **6**, 1024–1044.
4. N. Zheludev, The life and times of the LED – a 100-year history, *Nat. Photonics*, 2007, **1**, 189–192.
5. N. Stafford, LEDs to light up the world, *Chem. World*, 2010, 42–45.
6. F. D. Rosi and R. Braunstein, Semiconductive materials for infrared transmissive components, *United States Patent*, 2 977 477, 1961.

7. R. Braunstein and E. E. Loebner, Semiconductor device for generating modulated radiation, *United States Patent*, 3 102 201, 1963.

8. P-n junction. http//en.wikipedia.org/wiki/P%E2%80%93n_jumction.

9. N. Holonyak, Jr, Forward biased negative resistance semiconductor devices, *United States Patent*, 3 249 764, 1966.

10. J. S. Biard and G. E. Pittman, Semiconductor radiant diode, *United States Patent*, 3 293 513, 1966.

11. N. Holonyak, Jr, Use of metallic halide as a carrier gas in the vapour deposition of III-V compounds, *United States Patent*, 3 249 473, 1966.

12. W. O. Groves, A. H. Herzog and G. Craford, Process for the preparation of semiconductor materials and devices, *United States Patent*, 3 873 382, 1975.

13. M. G. Craford and D. L. Keune, Luminescent solid state status indicators, *United States Patent*, 3 964 039, 1976.

14. M. G. Craford and P. T. Bailey, Integrated semiconductor light-emitting display array, *United States Patent*, 3 947 840, 1976.

15. J. I. Pankove, Electroluminescent semiconductor device of GaN, *United States Patent*, 3 683 240, 1972.

16. S. Nakamura, N. Iwasa and M. Senoh, Method of manufacturing p-type compound semiconductor, *United States Patent*, 5 468 678, 1995.

17. S. Nagahama, M. Senoh and S. Nakamura, Nitride semiconductor light-emitting and light-receiving devices, *United States Patent*, 6 172 382, 2001.

18. S. Nakamura, S. Nagahama, N. Iwasa and H. Kiyoku, Nitride semiconductor light-emitting devices, *United States Patent*, 6 580 099, 2003.

19. Impact Lighting Inc, *Technical LEDs. LED colour chart*, Impact Lighting Inc., Orlando, Florida, USA.

20. J. Maguire and D. Segal, IP strategy plays key role in LED business development, *LEDs Magazine*, 2012, 57–60.

21. T. Hamada, Backlight and liquid crystal display device, *United States Patent*, 8 289 478, 2012.

22. T. Yoshida, T. Osaki, T. Kosaka and A. Ban, Lighting device, display device and television receiver, *United States Patent Application*, 2012/0262633, 2012.

23. M. R. Krames, G. O. Mueller, R. B. Mueller-Mach, H.-H. Bechtel and P. J. Schmidt, Wavelength conversion for producing white light from high power blue LED, *United States Patent Application*, 2010/0289044, 2010.

24. Y. Shimizu, K. Sakano, Y. Noguchi and T. Moriguchi, Light emitting device having a nitride compound semiconductor and a phosphor containing a garnet fluorescent material, *United States Patent*, 5 998 925, 1999.

25. T. Whitaker, Philips' LED licensing program attracts more licenses, *LEDs Magazine*, 2012.

26. M. L. Wach, Highly integrated system and method for optical communication, *United States Patent*, 8 611 756, 2013.

27. M. Kobayashi, Semiconductor laser and manufacturing process thereof, *United States Patent*, 8 111 728, 2012.

28. S. G. Park, M. H. Park, J. M. Lee, S. H. Oh and K. H. Kim, Method of wavelength tuning in a semiconductor tunable laser, *United States Patent*, 7 065 108, 2006.

29. Y. Fan, J. Han, A. V. Nurimikko, R. L. Gunshor and L. He, Group II-VI compound semiconductor light emitting devices and an ohmic contact therefor, *United States Patent*, 5 548 137, 1996.

30. W. H. Culver, Wedge side pumping for fibre laser at plurality of turns, *United States Patent*, 5 923 694, 1999.

31. P. D. Dapkus, Method and apparatus for long wavelength semiconductor lasers, *United States Patent*, 6 621 842, 2003.

32. S. Karakida, M. Miyashita and Y. Mihashi, Semiconductor laser device, *United States Patent*, 5 675 601, 1997.

33. K. Takagi, Semiconductor laser, *United States Patent*, 7 773 651, 2010.

34. M. A. Kneissl, Blue and green laser diodes with gallium nitride or indium gallium nitride cladding laser structure, *United States Patent*, 7 751 455, 2010.

35. S. R. Kurtz, R. M. Biefeld and A. A. Allerman, Infrared light sources with semimetal electron injection, *United States Patent*, 5 995 529, 1999.

36. Y. Kuromizu, Semiconductor light emitting device, *United States Patent*, 7 672 347, 2010.

37. A. Nakamura, K. Ikemoto and S. Takahashi, Semiconductor laser and semiconductor laser module, *United States Patent*, 6 873 636, 2005.

38. Y. Onishi, Semiconductor laser device and method for producing the same, *United States Patent*, 8 477 820, 2013.
39. T. Takiguchi and C. Watatani, Buried type semiconductor laser, *United States Patent*, 7 720 123, 2010.
40. T. Okumura, F. Konushi, T. Morioka and N. Matsumoto, Semiconductor laser device, *European Patent*, 0 566 411 B, 1998.
41. S. Akiba, M. Usami, Y. Matsushima, K. Sakai and K. Utaka, Quantum well structure and semiconductor device using the same, *European Patent*, 0 353 054 B, 1997.
42. P. Gao, Tunable laser, *United States Patent Application*, 2013/0022062, 2013.
43. L. P. O. Lundqvist, Tunable semiconductor laser with integrated wideband reflector, *United States Patent*, 6 822 980, 2004.
44. S. Taghavi-Larigani, J. J. Vanzyl and A. Yariv, Tunable semiconductor lasers, *United States Patent*, 7 027 476, 2006.
45. T. B. Mason, G. Fish and L. Coldren, Method of making a tunable laser source with integrated optical amplifier, *United States Patent*, 6 654 400, 2003.
46. D. Feklistov, Diode pumped laser, *United States Patent Application*, 2008/0253419, 2008.

CHAPTER 3

Quantum Dots

3.1 INTRODUCTION

The scientific study of colloids dates back to 1845 when Selmi prepared silver chloride dispersions (sols) followed by Prussian blue sols in 1847.[1,2] Faraday observed light-scattering from ruby-coloured gold sols made by reduction of gold chloride with phosphorus together with the effect of salt on their stability and colour.[3] However, the word colloid was first used by Graham in 1861 to describe 'glue-like' material prepared by dialysis of silicic acid made by acidifying silicate solutions and also organic species such as gums, caramel, tannin and albumin.[4] Nowadays, colloidal systems are defined as comprising a disperse phase with at least one dimension between 1 nm and 1 μm in a dispersion medium. For sols the colloidal dimension refers to particle diameter, whereas for macroscopic colloidal systems such as foams it refers to film thickness.[5]

In recent years nanotechnology has attracted much attention and is often promoted as promising a future based on materials with properties that were previously unattainable. Discussions on nanotechnology often refer to small particles described as nanoparticles, nanocrystals, nanomaterials and, of interest here, quantum dots. Surprisingly, very small particles that fall within the domain of nanotechnology also fall within the colloidal

Exploring Materials through Patent Information
By David Segal
© David Segal, 2015
Published by the Royal Society of Chemistry, www.rsc.org

range and as described here, colloidal dispersions have an established and distinguished history. The excellent book by Ozin *et al.*[6] gives an outstanding account of nanomaterials and applications and it has an extensive set of references based on journal references. In keeping with the approach set out for this book an account of quantum dots is given here with, where possible, references from the patent literature in order to highlight the chronological development of the subject.

3.2 QUANTUM DOTS: STRUCTURE

Electrons confined in a potential well whose size is comparable to that of the de Broglie wavelength of an electron behave differently from those in free space. In particular, the electronic and optical properties of metal and semiconductors change sharply as the particle sizes are reduced to the nanometre size range.[7,8] The restriction of the electronic wave function to small regions of space in a material is known as quantum confinement.[7] An extremely thin film structure or one-dimensional quantum well has dimensions equal to the de Broglie wavelength of an electron when confinement is in one dimension,[9] whereas a two-dimensional quantum well is referred to as a quantum wire or nanowire. Quantum dots (or quantum boxes) are semiconductor structures where the confinement is in three dimensions and whose linear dimensions in three directions are less than the de Broglie wavelength of the electrons. A quantum dot has a potential well that allows a quantum mechanical standing wave to be established for an electron. The difference in the dimensionality of the electrons results in a large difference in their density of states and energy levels. Quantum dots have discrete energy levels rather like atoms and molecules.

3.3 QUANTUM DOTS: SYNTHESIS

Many techniques have been used to prepare quantum dots and examples of these are given in this section. An aluminium foil was anodized in an acid bath, producing an aluminium oxide film containing uniform diameter micropores in the range 1–500 nm with a preferred diameter in the range 10–100 nm.[7] A metal capable of forming a semiconductor, Al, Cd, In, Ga or Zn was deposited electrolytically in the pores and deposited material was etched with a solution of 0.47 M phosphoric acid and 0.2 M chromic acid to produce rods of uniform height. Part of the oxide film surrounding the Cd rods was etched away to leave electrical contact for the rods. Exposed cadmium metal was reacted with S vapour at 300 °C to form CdS rods with diameters in the range 60–100 nm. Electron beam diffraction has been used to produce an array of quantum dots. An AlGaAs layer was first formed on either an AlGaAs or GaAs substrate after which a GaAs layer was epitaxially grown on the wafer surface until its thickness corresponds to the height of the quantum dots.[10] An electron beam was diffracted in a chamber as it passed through a GaAs membrane, after which it irradiated the wafer, resulting in transfer of the diffraction pattern from the membrane onto the upper surface of the GaAs layer. Carbon was introduced into the electron beam chamber and adhered to the regions on the GaAs layer, producing a carbon mask structure. Reactive ion etching (RIE) was then used to remove the GaAs layer not covered by the carbon mask, after which the latter was removed and an AlGaAs layer deposited on exposed portions of the AlGaAs layer by epitaxial growth. Quantum dots were formed in the positions marked out with the carbon mask pattern and each of the dots was insulated from one another by the AlGaAs layer. Spacing between the dots was 5–20 nm and the dot diameter around 10 nm.

Ion implantation has also been used for fabrication of quantum dots and wires.[11] Hence an epitaxial GaAs/Al_xGa_{1-x}As heterostructure or $Si_{1-y}Ge_y$/Si hetero-structure was grown on a single crystal GaAs or Si substrate where $0.1 < x < 0.7$ and $0.05 < y < 0.5$. Metal-organic chemical vapour deposition (MOCVD), molecular beam epitaxy (MBE) or focused ion beams (FIB) were used in the deposition process. The narrower band gap semiconductor, GaAs or $Si_{1-y}Ge_y$ (the channel layer) was grown, in the case of GaAs, to a thickness of up to 1000 nm on top of a GaAs buffer with thickness between 200 and 300 nm. FIB was used in a vacuum chamber to implant a dot-like or wire-like pattern of dopant into the wider bandgap layer forming an n-type electron supply layer or p-type electron hole supply layer and a zero- or one-dimensional electron gas or electron hole gas in the channel layer. Ions from the FIB apparatus do not extend to the channel layer but extend close enough to allow a sufficient supply of electrons and holes into the channel layer. Damage to the channel layer is avoided due to the absence of impurity scattering and ion impurities in the channel as well as the absence of chemical etching processes.

Quantum dot structures have been fabricated by thermal etching.[12] For example, a single quantum well formed by an $In_{0.25}Ga_{0.75}As$ layer of 5 nm thickness was grown on a GaAs substrate. An AlAs masking layer with a thickness of 0.5 atomic layers was overlaid on the $In_{0.25}Ga_{0.75}As$ layer by molecular beam epitaxy. This masking layer did not cover the surface uniformly. When the wafer was heated at 670 °C for 3 minutes, regions of $In_{0.25}Ga_{0.75}As$ not covered by the masking layer were evaporated off. After thermal etching, quantum dots were covered with a layer of material having a bandgap energy greater than the bandgap of the quantum well layer.

The difference in lattice spacing and resulting strain between semiconducting layers has been used to produce quantum dots.[13] An epitaxial layer with a lattice spacing greater than that for the substrate material places the layer under compression whereas a layer is in tension when its lattice spacing is less than that for the underlying substrate. Molecular beam epitaxy was used to produce a partial layer of InGaAs on a GaAs substrate by exposure of the latter to AsH_3. The surface texture of the InGaAs layer was monitored continually by electron beam diffraction as it grew on the substrate to observe how the surface texture changed with composition. The mismatch in lattice spacing between epitaxial layer and substrate leads to formation of three-dimensional islands of quantum dots on the surface that are characterised by the presence of a 'spotty' electron diffraction pattern when the substrate was held at 530 °C. Examples of compositions for the quantum dots were $In_{1-x}Ga_xAs/GaAs$ and $In_{1-x}Al_xAs/Al_yGa_{1-y}As$ where $0 < x < 1$ and $0 < y < 1$ as well as $In_{0.25}Ga_{0.75}As$. The InAlAs system was luminescent at visible wavelengths while the InGaAs system luminesced in the near infra-red. Diameters of the quantum dots were in the range 14–30 nm.

Methods that have been described here for synthesis of quantum dots are based on gas phase reactants. Quantum dots fall within systems with a colloidal dimension and in recent years wet chemical methods have attracted attention for synthesis of fine particles, particularly ceramic particles and synthesis of quantum dots, for example aerosol systems.[5,6,14] A solution of 10^{-3} M cadmium nitrate containing sodium hexametaphosphate was first diluted with ethyl alcohol. Aerosol droplets could be produced with diameters 0.1–10 μm using an electrostatic spray nozzle but experimental conditions were arranged for droplets with diameters between 0.1 and 1 μm in a gaseous atmosphere containing 5% H_2S at 573 °C. Ethyl alcohol and water evaporated from the droplets leaving CdS particles with a mean diameter of 6 nm and inorganic polyphosphate coating derived from the hexametaphosphate. Other non-oxide particles based on Zn, Pb, Sn, Hg, Al, Ga, In, Ti, Si and Ca could be prepared using a gaseous atmosphere containing H_2S, H_2Se, H_2Te, PH_3, AsH_3 and SbH_3.

Hydrothermal processing routes to ceramic powders involve heating reactants such as metal salts, oxides or hydroxides as a solution or suspension in a liquid, usually water, at elevated temperature and pressure. For example, CdS was produced by ageing a mixture of 0.0012 M Cd $(NO_3)_2$, 0.24 M HNO_3 and 0.005 M thioacetamide at 572 °C for 14 h.[15] In a recent synthesis,[16] a mixture of Group II and Group III precursors in a solvent was heated to between 200 °C and 350 °C. For example, indium oleate and zinc oleate solutions were heated in 1-octadecene at 120 °C under N_2 and then, when the atmosphere was oxygen-free, the temperature was raised to 300 °C. Then a mixture of Group V and Group VI precursors was added. In this synthesis a mixture of tributyl phosphine sulphide and tri-methylsilyl phosphine was injected and the reaction temperature maintained at 300 °C. Particles of the quantum dot had an InP core and InZnS shell. For other syntheses the Group II element could be Zn, Cd, Hg and the Group III element Al, Ga, In. The Group V precursor could be an alkyl phosphine or tris (dialkyl-amino phosphine) as examples and the Group VI precursor derived from S, Se or Te, for example an alkylamino sulphide or alkylamino selenide. Hydrothermal synthesis of quantum dots with composition CdS, CdSe and CdTe as well as the core-shell structures CdSe/ZnS, CdSe/CdS and CdSe/CdS/ZnS has been described.[17] A key feature is use of the microwave assisted processing and synthesis (MAPS) route that enabled rapid

heating and stable reaction temperatures for the precursor solutions. In the case of CdSe, a solution of cadmium perchlorate
in ultrapure water was mixed with sodium citrate and N,
N-dimethylselenourea. This mixture was deaerated, sealed in a
glass vial and hydrothermally processed in a microwave reactor
between 60 °C and 180 °C for 5 minutes. Microwaves promoted
aqueous nucleation and growth of CdSe by better control of the
reaction conditions. Core-shell dots were also prepared. For
example, as-prepared citrate-capped CdS nanoparticles were
added to thioacetamide and zinc perchlorate solution and then
hydrothermally-processed for 2 minutes at 120 °C; the zinc
source, $Zn(ClO_4)_2$ promoted epitaxial growth of a ZnS shell.

Hydrothermal processing has been applied to nitride-
containing quantum dots.[18] Hence for InN, a mixture of indium
iodide, sodium amide (a nitrogen source), hexadecane thiol
(a capping agent with an electron-donating group), zinc stearate
(a capping agent with an electron-accepting group) and a solvent
diphenyl ether were heated at 225 °C. Zinc stearate, the capping
agent with an electron-accepting group increased the photo-
luminescent quantum yield of the nitride nanoparticle. For
InGaN, gallium iodide, indium iodide, hexadecane thiol, zinc
stearate, sodium amide and 1-octadecene were heated to and
maintained at 225 °C for periods up to 60 minutes. For core-
shell InN-ZnS nanocrystals, indium iodide, sodium amide,
hexadecane thiol, zinc stearate and 1-octadecene were heated to
250 °C and kept at that temperature for 30 minutes. After cooling, insoluble material was removed by centrifugation and the
residual solution treated with zinc diethyl dithiocarbamate
at 175 °C for 60 minutes, where the latter reacted to form the
ZnS shell. A dispersion of InN-ZnS core-shell nanocrystals
was obtained on cooling the reaction mixture and removing
insoluble material by centrifugation.

3.4 QUANTUM DOTS: PROPERTIES AND APPLICATIONS

The interest in quantum dots resides in the fact that the electrical and optical properties of materials and semiconductors
changes as the size range of the material approaches the nanometre region, that is less than around 10 nm. For example,
Figure 3.1 compares the absorption spectrum of CdS quantum

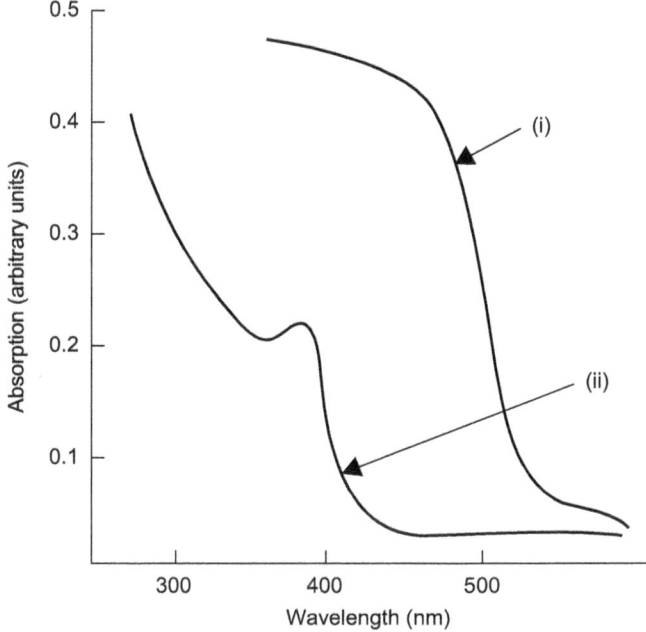

Figure 3.1 Absorption spectrum of (i) macrocrystalline cadmium sulphide and (ii) cadmium sulphide quantum dots.[14]

dots and macroscopic CdS and shows their clear differences.[14] Quantum dots are semiconductor phosphors and emit wavelengths longer than that of the exciting radiation, that is, they exhibit fluorescence. Thus, the core-shell dot with an InP core and InZnS shell exhibited a maximum emission wavelength between 508 and 536 nm depending on the concentration of precursors used in its synthesis when excited by light with a wavelength of 400 nm.[16]

A single chip type white light LED has been developed by Yang *et al.*[19] as shown schematically in Figure 3.2. In this configuration, mixing light from a red light-emitting quantum well with green and blue light emitted from green and blue light-emitting quantum dots produced white light. The mixing process avoided problems of colour shift in conventional LEDs while similar lattice structures of the components in the chip reduce formation of dislocations and enhance the light-emitting efficiency. Potential uses of quantum dots in LEDs are not surprising

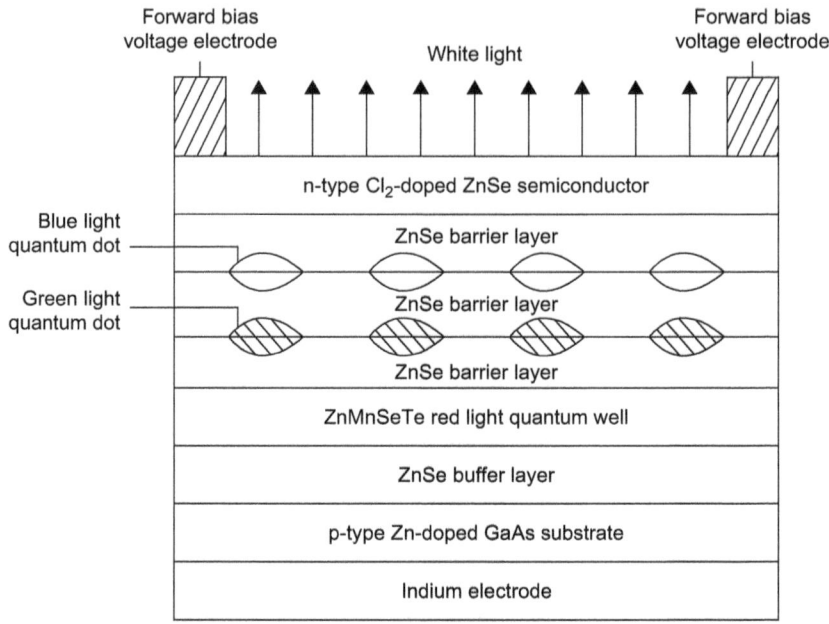

Figure 3.2   Schematic diagram of a single chip type white LED device.[19]

because LEDs are becoming ubiquitous in modern display technology.[20] More than 30×10^9 chips are produced each year and applications continue to grow. LEDs emit monochromatic light of a frequency corresponding to the semiconductor bandgap in the device. Production of LEDs with a particular pure single frequency can be difficult as tight control of semiconductor chemistry is required. There is thus a continued demand for LEDs with improved light-emitting properties. A primary light source was obtained using a blue diode based on a GaN semiconductor with a luminescence peak at 450 nm.[20] This light passed through a layer of core-shell CdSe-ZnS quantum dots embedded in a polymer matrix and red secondary light was emitted. Primary light not adsorbed by the nanoparticles passes with the red secondary light into an embedded layer of green light-emitting quantum dots. Red, green and blue components were then combined to form white light. Amplitudes of red, green and blue components were varied by changing the layer thickness and the number density of quantum dot particles in the layer to produce specific colours. Quantum dots with a polydisperse size distribution allow light of any colour to be produced. For example, 5.5 nm CdSe nanoparticles were red emitters, 4.0 nm nanoparticles were green emitters while 2.3 nm nanoparticles were blue emitters. Note that as the particle size decreases there is a shift to shorter wavelengths for emitted light corresponding to an increase in bandgap energy with decrease in particle size.

Schematic diagrams of a quantum dot LED and a quantum well LED are shown in Figure 3.3.[21] In the quantum dot LED an insulator layer is sandwiched between n-type and p-type semiconductor layers and nanoholes with a size range 1–100 nm extended throughout the insulator layer. Nanoholes were filled with a material constituting a quantum dot, such as InGaN, InGaAs and InGaP. The upper surfaces of the quantum dots are in contact with p-type semiconductor layers and the lower

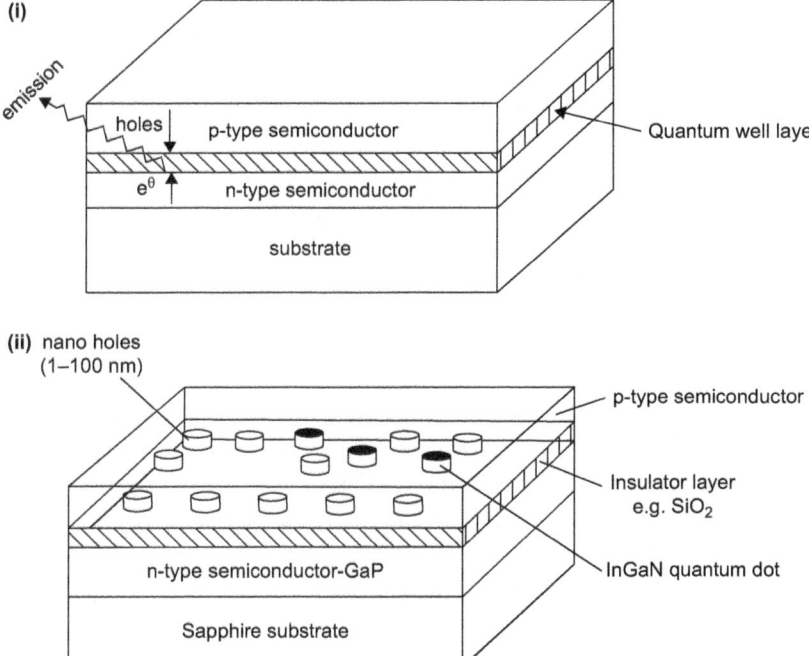

Figure 3.3 Schematic diagram of (i) LED with quantum well light-emitting layer and (ii) quantum dot LED.[21]

surfaces of the quantum dots are in contact with n-type semi-conductor layers. Under forward bias electrons and holes re-combine so that light is emitted from the quantum dots and the characteristics of the LED device are controlled by the size and number density of the dots in the insulator layer. The internal quantum efficiency is higher in the quantum dot LED than in the conventional LED with a quantum well layer with an associated lower power consumption.

Nanophosphors, that is, fluorescent material with diameters representative of quantum dots, namely 2–10 nm (10–50 atoms in diameter) have been considered as secondary light sources in an LED backlight for a LCD panel.[22] When narrow passband and narrow absorption band filters with a full width at half max-imum transmissivity (FWHM) of 100 nm or below were com-bined with light emissions from the quantum dots a high efficiency and high NTSC ratio (named after US National Tele-vision System Committee) was produced. This approach helps to avoid colour shifts that can occur if relative degradation of three phosphors (red, green, blue) used in backlights are different. Flexible substrates containing a light-emitting layer of quantum dots have been suggested for use as advertising hoardings.[23] The light-emitting layer could be painted or coated with a quantum dot ink. An external light source, visible or ultra-violet, was used to excite quantum dots while the brightness of the display was controlled by the number of dots contained in the unit area of the surface.

The demand for renewable energy solutions to worldwide concerns on global warming and energy supplies has generated much research and development on solar or photovoltaic cells. Photons emitted from the sun are absorbed by semiconductors and excite electrons from the semiconductor's valency band across the bandgap into the conduction band. Electron-hole pairs are separated in the solar cell to generate a photocurrent and photovoltaic effect. Photons impinging with energies greater than or equal to the bandgap energy are absorbed but no ab-sorption takes place when the photon energy is less than the bandgap energy.

Figure 3.4 is a schematic diagram of a quantum dot solar cell.[24] A glass substrate with a transparent electrode coating such as indium tin oxide (ITO) transmits sunlight into a layer of

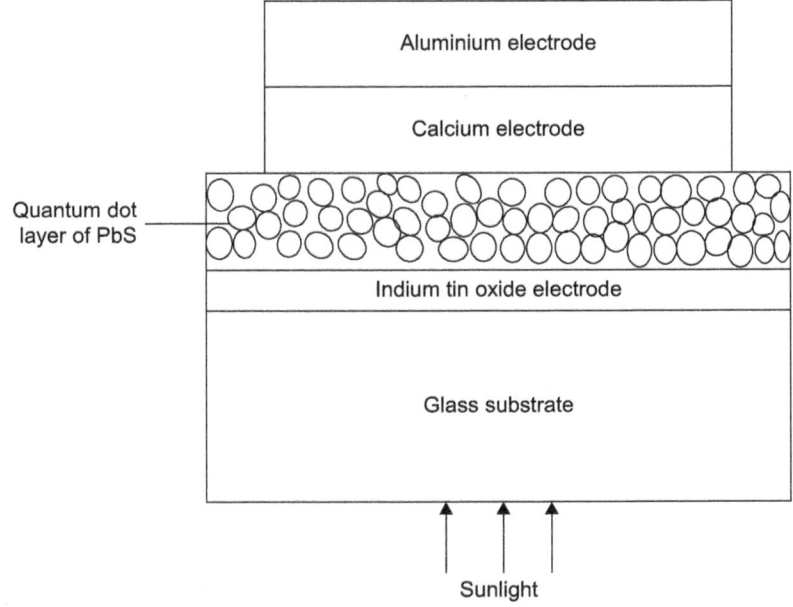

Figure 3.4 Schematic diagram of a quantum dot solar cell.[24]

quantum dots, PbS or PbSe or CdS as examples, although core-shell and ternary nanocrystals can be used, hence lead selenide cores with lead sulphide shells or PbSSe (lead selenide sulphide). Quantum dots were deposited as a layer onto the ITO electrode by dip-coating the latter in a dispersion of PbS nanoparticles (absorption peak at 1800 nm) in ethanedithiol after which the substrate was removed and dried. This dip-coating process was repeated up to 20 times. A Ca electrode was then deposited to a thickness of 20 nm and protected from degradation by water vapour with an Al layer around 100 nm thick. Instead of dip-coating a layer of quantum dots onto a substrate, epitaxial growth of nanoparticles has been used as an alternative approach.[25] For example, InAs quantum dots were intercalated with InGaAs layers on an InP substrate to tailor the semiconductor bandgap energy and improve the conversion efficiency of sunlight to electrical energy in the photovoltaic cell.

Coated nanoparticles, particularly coatings of Cu, In, Ga, S or Se have been used as absorber layers for low-cost deposition routes to flexible solar cell substrates using printing or web-coating techniques.[26] For example, Cu or Cu/Ga nano-particles with an average particle size of 20 nm were coated, by electroless plating, with an In layer around 10–50 nm thick. These quantum dots could be used in a colloidal dispersion (sol) to form inks, paint or paste for depositing a film on a flexible substrate in a roll-to-roll manner using a web-coating system. The substrate could be metal foil (*e.g.* Al) or a polymer such as polyimide or polyethersulphone while sulphur or selenium could be incorporated into the film as sulphide or selenide by exposing the film during annealing at 200–600 °C to H_2S or H_2Se. A further example of flexible solar cells has been described by Wang *et al.*[27] An array of quantum dots, for example CdSe, are deposited on a patterned TiO_2 layer on a flexible electrode substrate (*e.g.* Ti foil or poly(ethylene terephthalate)).

Quantum dots are not restricted to lighting and photovoltaic applications. For example, a quantum dot laser was constructed by embedding 5 nm diameter CdSe nanoparticles that fluoresce at 550 nm in a polymethylmethacrylate matrix and pumped with an argon ion laser operating at 488 nm to induce a population inversion in the quantum dots.[28] Terahertz radiation has a wavelength between the infra-red region and the microwave region of the electromagnetic spectrum, in particular the region between 0.03 mm and 3 mm.[29] This range of wavelengths has potential applications in imaging as it can penetrate organic material, fabrics and plastics without the damage associated with ionising radiation such as X-rays and can be used in surveillance such as airport screening to identify concealed weapons. Terahertz radiation was generated by subjecting a photoconductor with a quantum dot layer containing PbS or CdTe or CdSe/ZnS or ZnCdSe/ZnS to incident laser radiation.

Quantum dots are finding biomedical applications where changes in luminescence can be used as a signal in receptor-ligand binding assays and thus can be used to detect an analyte.[30]

3.5 SUMMARY

Quantum dots are semiconductors whose electrical and optical properties are determined by quantum confinement effects due to their particle size, which is typically around 10 nm. Properties are dependent not just on material composition but also on particle size. Quantum dots behave as phosphors and fluoresce, emitting wavelengths larger than that of the exciting radiation. There is increasing interest in the application of quantum dots for displays in electronic devices and for use in solar cells.

REFERENCES

1. F. Selmi, Studies on the demulsion of silver chloride, *Nuovi Ann. Sci. Nat. Bologna*, 1845, **2**(4), 146.
2. F. Selmi, A study of the pseudo-solutions of Prussian blue and of the influence of salts in destroying them, *Nuovi Ann. Sci. Nat. Bologna*, 1847, **2**(8), 401.
3. M. Faraday, On the experimental relations of gold (and other metals) to light, *Philos. Trans. R. Soc. London*, 1857, **147**, 145.
4. T. Graham, Liquid diffusion applied to analysis, *Philos. Trans. R. Soc. London*, 1861, **151**, 183–224.
5. D. Segal, *Chemical synthesis of advanced ceramic materials*, Cambridge University Press, 1989.
6. G. A. Ozin, A. C. Arsenault and L. Cademartiri, *Nanochemistry: A chemical approach to nanomaterials*, Royal Society of Chemistry, Cambridge, 2nd edn, 2009.
7. M. Moskovits, Process for manufacture of quantum dot and quantum wire semiconductors, *United States Patent*, 5 202 290, 1993.
8. T. Fukazawa and H. Munekata, Fabrication method for quantum devices in compound semiconductor layers, *United States Patent*, 5 281 543, 1994.
9. Y. Kato, Method of fabricating a semiconductor device using quantum dots or wires, *United States Patent*, 5 532 184, 1996.
10. R. Ugajin, Method for forming quantum dots, *United States Patent*, 5 229 320, 1993.
11. Y. Kato, Semiconductor device having a channel for a zero- or one-dimensional carrier gas, *United States Patent*, 5 479 027, 1995.
12. D.-C. Liu and C.-P. Lee, Method of fabricating quantum dot structures, *United States Patent*, 5 482 890, 1996.
13. P. Petroff, D. Leonard and M. Krishnamurthy, Quantum dot fabrication process using strained epitaxial growth, *United States Patent*, 5 614 435, 1997.

14. P. J. Dobson, O. V. Salata, P. J. Hull and J. L. Hutchison, Making particles of uniform size, *United States Patent*, 5 906 670, 1999.
15. E. Matijevic, Monodispersed colloidal metal oxides, sulphides and phosphates, in *Ultrastructure Processing of Ceramics, Glasses and Composites*, ed. L. L. Hench and D. R. Ulrich, John Wiley & Sons, New York, 1984, 334–52.
16. J. H. Kang, J. Shin, J. B. Park, D-Hoon Lee, M. Nam, K. Char, S. Lee, W. Bae, J. Lim and J. Jung, Method of manufacturing quantum dots, *European Patent Application*, EP 2368964A2, 2011.
17. M. Maye, Greener synthesis of nanoparticles using fine tuned hydrothermal routes, *United States Patent Application*, 2012/0103789, 2012.
18. P. N. Taylor and J. Heffernan, Fabrication of nitride nano-particles, *United States Patent Application*, 2012/0018774, 2012.
19. C.-S. Yang, C.-S. Wu, W.-C. Chou, M.-T. Chiang, C.-N. Mo, C.-W. Luo and L.-K. Huang, Single chip type white LED device, *United States Patent Application*, 2011/0108797, 2011.
20. M. G. Bawendi, J. Heine, K. F. Jensen, J. N. Miller and R. L. Moon, Quantum dot white and colored light-emitting devices, *United States Patent*, 8 174 181, 2012.
21. S. C. Choi, LED and fabrication method thereof, *United States Patent*, 8 017 931, 2011.
22. R. Roshan, P. N. Taylor and D. J. Montgomery, Display, *United States Patent*, 7 686 493, 2010.
23. J. W. Song, Display device using quantum dot and fabrication method thereof, *International Patent Application*, WO2011/049311, 2011.
24. A. J. Nozik, M. Beard, M. D. Law and J. M. Luther, Solar cells based on quantum dot or colloidal nanocrystal films, *United States Patent Application*, 2011/0146766, 2011.
25. S. Fafard, Solar cell with epitaxially grown quantum dot material, *United States Patent*, 7 863 516, 2011.
26. B. M. Sager, D. Yu and M. R. Robinson, Coated nanoparticles and quantum dots for solution-based fabrication of photovoltaic cells, *United States Patent*, 8 193 442, 2012.
27. M. Wang, L. Zhao, Z. Zheng and A. Liu, Flexible quantum dot sensitized solar cells, *European Patent Application*, EP2442325A, 2012.

28. F. Hakimi, M. G. Bawendi, R. Tumminelli and J. R. Haavisto, Quantum dot laser. *United States Patent,* 5 260 957, 1993.

29. E. Rafailov, N. Bazieva, N. Daghestani and D. Turchinovitch, Terahertz source with quantum dot material, *International Patent Application*, WO2012/110759, 2012.

30. F. M. Raymo, M. Tomasulo and I. Yildiz, Mechanism to signal receptor-ligand interactions with luminescent quantum dots, *United States Patent*, 8 198 099, 2012.

CHAPTER 4

Organic Light-emitting Diodes

4.1 INTRODUCTION

A revolution is taking place in lighting technology and this area is one of rapid technological change. Light-emitting diodes were discussed in Chapter 2 and represent efforts in research, development and commercial exploitation for over a hundred years, embracing advances in chemistry, physics and materials science. Applications for LEDs continue to expand and more than 30 billion LED chips are produced worldwide each year. In addition, quantum dots, whose properties are based on the quantum confinement effect in semiconductors, exhibit fluorescence when excited by an external light source. Quantum dots can be viewed as colloidal systems, as explained in Chapter 3, and the scientific study of such systems dates back to 1845. Organic light-emitting diodes (OLEDs) are another type of LED and have a more recent history, with origins in the 1950s. The demand for cheap, compact and low-power consuming light sources for displays in the increasing range of consumer products and electronic devices helps to drive research and development in light-emitting diodes, quantum dots and OLEDs. Activity on the latter is described here from the perspective, when available, of published patent literature.

Exploring Materials through Patent Information
By David Segal
© David Segal, 2015
Published by the Royal Society of Chemistry, www.rsc.org

4.2 THE EARLY YEARS

In early work, Bernanose detected electroluminescence in organic compounds including gonacrin, brilliant acridine orange E and carbazole when adsorbed into a cellulose film mounted as a dielectric in a condenser in which an alternating potential was applied.[1] Gurnee and Fernandez identified organic electroluminescent phosphors from a host cyclic organic compound containing a conjugated structure.[2] For example, the host material anthracene was doped with 0.1 weight % of tetracene and about 1 weight % of a finely divided electrical conductor such as carbon powder. The resulting paste was contained as a 25 μm layer between flat electrodes, one of which was transparent. Green light was emitted from the region of the paste on application of an alternating potential of 800 volts. Other materials which exhibited electroluminescence included naphthalene, terphenyls and phenanthrene.

Kallmann and Pope contained a layer of anthracene crystals, 10 μm thick, between two liquid electrodes, a NaCl solution acting as anode and ceric sulphate solution, $Ce(SO_4)_2$ as cathode.[3] The electrodes injected electrons and holes into the anthracene layer, increasing its electrical conductivity, but it was not observed because currents in the anthracene crystals were too low; other compounds evaluated included naphthalene and molecular structures with conjugated double bonds such as carotenoids. Williams constructed a device (Figure 4.1) that monitored electroluminescence from an active organic material

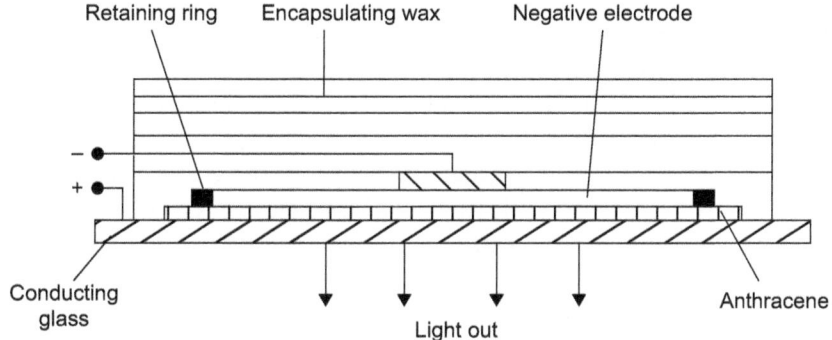

Figure 4.1 Device for showing electroluminescence from anthracene.[4]

such as anthracene.[4] An anthracene layer between 1 μm and 200 μm in thickness was in contact with a conducting glass electrode. The other electrode consists of solid material that can produce stable ions (*e.g.* anthracene anions) of a hydrocarbon similar to that used as the active material. A retaining ring and wax were used to give mechanical stability to the device. Electrons were injected from one electrode and recombined, in the anthracene layer, with holes injected from the other electrode. Electroluminescence was observed in the ultra-violet region for naphthalene at 350–400 nm, between 410 and 450 nm in the blue range for anthracene, whereas anthracene doped with tetracene gave green-blue emissions around 480–500 nm. However, the device had high power consumption and low luminescence.

Heeger *et al.* developed a series of p-type electrically conducting doped-polycrystalline films whose room temperature electrical conductivity could have values in the range characteristic of semiconductors and in the range characteristic of metallic behaviour.[5] Flexible polycrystalline films of polyacetylene with thicknesses between 0.1 μm and 1000 μm were prepared by polymerisation of acetylene monomer in the presence of titanium butoxide, $Ti(OC_4H_9)_4$/triethyl aluminium $Al(C_2H_5)_3$ catalyst. The room temperature electrical conductivity of as-prepared polyacetylene was in the range 10^{-9} to 10^{-5} $ohm^{-1}cm^{-1}$ depending on whether cis or trans isomers were produced in the polymerisation process. Polycrystalline polyacetylene films were exposed to vapour of electron-acceptor dopants for periods of up to 20 hours. These dopants include HBr, Cl_2, Br_2 and arsenic pentafluoride, AsF_5. Examples of the electrical conductivity of doped polyacetylene compositions are shown in Table 1 for cis and trans isomeric forms of the polymer.

For organic electroluminescent devices to compete with inorganic counterparts their power conversion efficiencies have to be of a sufficient value for acceptable brightness. Van Slyke and Tang showed how to obtain acceptable power conversion efficiencies defined as the ratio of power output (watts) to power input (watts) at low driving voltages.[6] Figure 4.2 shows a schematic diagram for a device that had a power conversion efficiency of at least 9×10^{-5} w/w at a driving voltage less than 25 volts. In this device the combined thickness of the

Table 4.1 Compositions and room temperature electrical conductivity for doped cis- and trans-polyacetylenes.[5]

Doped polyacetylene film composition	Conductivity at 25 °C (ohm^{-1}cm^{-1})
trans-$[(CH)(HBr)_{0.04}]_x$	7×10^{-4}
trans-$(CHBr_{0.23})_x$	4×10^{-1}
cis-$[CH(ICl)_{0.14}]_x$	5.0×10^1
trans-$(CHI_{0.22})_x$	3.0×10^1
cis-$[CH(IBr)_{0.15}]_x$	4.0×10^2
cis-$[CH(AsF_5)_{0.1}]_x$	8.8×10^2

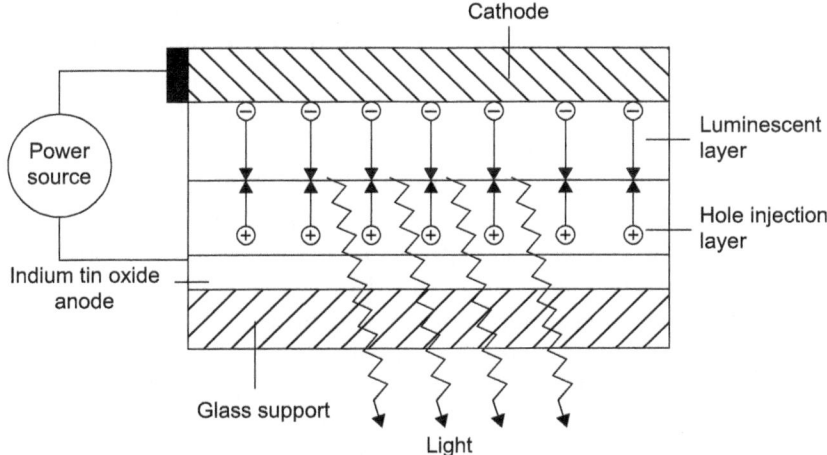

Figure 4.2 Schematic diagram of device with power conversion efficiency greater than 9×10^{-5} w/w.[6]

luminescent and hole injection layers was less than 1 μm. The hole injection layer consisted of 1,1-bis-(4-di-p-tolylaminophenyl)cyclohexane whereas the luminescent layer consists of an electron-transporting compound such as 4,4′-Bis(5,7-di-t-pentyl-2-benzoxazolyl)-stilbene and light emitted by electroluminescence had a wavelength greater than 400 nm.

Work carried out by Van Slyke and Tang concentrated on small molecule light-emitting diodes (SMOLEDs).[6] Activities on SMOLEDs have continued with (i) use of polyaromatic amines as additives to the hole transport layer for improving thermal stability over a range of operating temperatures,[7] (ii) fluorescent material that emits light in response to electron-hole re-combination,[8] and (iii) green electroluminescent devices.[9]

4.3 PIONEERING WORK ON CONJUGATED POLYMERS

The foundations for a new generation of flat panel displays based on electroluminescence from conjugated polymers was established by Friend *et al.*[10] The conjugated polymer is poly(p-phenylenevinylene) (PPV), whose molecular structure is shown schematically in Figure 4.3. The authors describe a conjugated polymer as 'a polymer which possesses a delocalized Π-electron system along the polymer backbone; the delocalized Π-electron system confers semiconducting properties to the polymer and gives it the ability to support positive and negative charge carriers with high mobilities along the polymer chain'. Polymer light-emitting diodes are referred to as PLEDs.

An electroluminescent device was constructed by initially forming a charge injecting contact layer of aluminium about 20 nm in thickness on a glass substrate. The aluminium charge injecting contact layer was exposed to the air to allow formation of a thin surface layer of aluminium oxide that formed the electron injecting contact layer. A solution of the precursor to PPV was spin-coated onto the aluminium oxide film and converted to PPV after heating at 300 °C in a vacuum. The PPV film had a thickness in the range 100–300 nm. A second charge injecting contact layer was formed by evaporation of Au or Al onto the PPV film. This layer, with thickness in the range 20–30 nm, formed the hole-injecting contact layer. Materials suitable for use as an electron-injecting contact layer have a low work function relative to the electroluminescent layer and include n-doped silicon (amorphous or crystalline), silicon with an oxide coating,

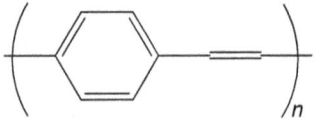

Figure 4.3 Schematic diagram showing the molecular structure of poly(p-phenylenevinylene); n, the degree of polymerisation, is not specified.[10]

alkali and alkaline-earth metals either pure or alloyed with other metals such as silver. Materials suitable for use as a hole-injecting contact layer because they have a high work function relative to the electroluminescent layer include indium tin oxide, platinum, nickel, palladium and graphite. The electroluminescent spectrum covered the spectral range 500–690 nm when a voltage was applied to the device.

Electroluminescence in inorganic semiconductors can be considered in terms of valence and conduction bands. When an OLED is subjected to forward bias, injected electrons and holes can recombine in the organic layer and emit light with a wavelength depending on the properties of the organic material. The concept of energetic bands with highly delocalized wave functions is not applicable to OLEDs.[11] It is customary to use the term 'HOMO level' (highest occupied molecular orbital level) and 'LUMO level' (lowest unoccupied molecular orbital level). Electron-hole pairs form an exciton which on decay exhibits electroluminescence and this is shown schematically in Figure 4.4.

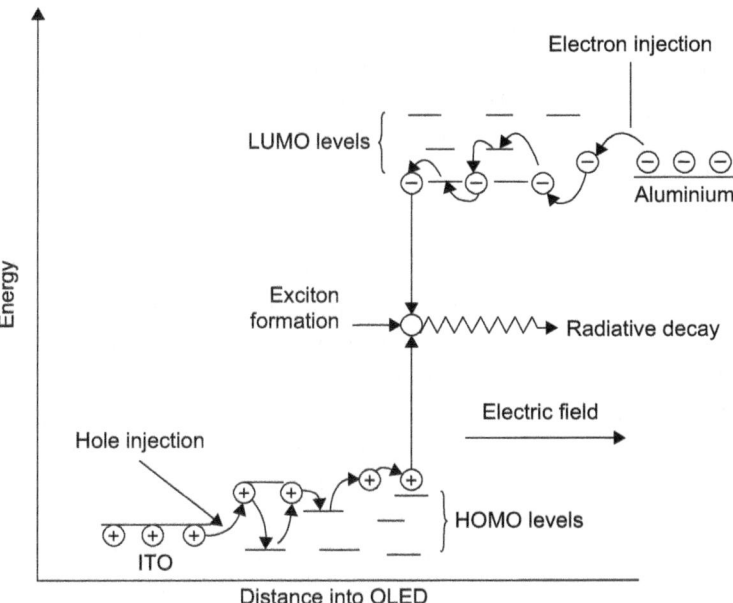

Figure 4.4 Schematic diagram showing electron and hole transport in an OLED.[11]

4.4   OLEDS: APPLICATIONS

It has been estimated that over 20% of the total U.S. electric energy production is consumed for lighting applications and that significant savings can be made by the introduction of new, higher performance illumination sources.[12] Organic light-emitting diodes as well as SMOLEDs are becoming increasingly desirable for new lighting sources. Precursors used to make OLEDs are relatively inexpensive and have the potential for cost savings over inorganic LEDs. Inherent properties of organic materials such as their flexibility make them suitable for fabrication on a flexible substrate. Also, the wavelength at which an organic emissive layer emits light may be tuned by addition of dopants while it may be more difficult to tune the inorganic materials used in LEDs.

Developments in OLED technology are taking place on several fronts. For example, organic phosphorescent materials emit light from triplet states,[13] whereas conventional OLEDs use molecules that emit light from their singlet states by fluorescence.[8] OLEDs that incorporate phosphorescent materials are referred to as PHOLEDs. An example of a green emissive phosphorescent material is tris(2-phenylpyridine) iridium, referred to as Ir(ppy)$_3$. Phosphorescent emitters are potentially more efficient than fluorescent emitters. Hence 65% of emitted light in a PHOLED lighting panel was traceable back to the phosphorescent material.

OLED and PHOLED display devices can have complex structures as shown in Figure 4.5.[14] The electron injection layer which facilitates electron injection into an emission layer may be formed of tris(8-quinolinolate) aluminium while the electron transport layer that facilitates electron transport into the emission layer may be formed of a polymer such as 2-(4-biphenyl)-5-(4-tert-butylphenyl)-1,3,4-oxadiazole. The hole blocking layer prevents diffusion of excitons generated from electron-hole recombination and bathocuproin can be used for this purpose. The emission layer can contain a fluorescent or phosphorescent material and an organic metal complex can be used for the latter. The electron blocking layer prevents diffusion of excitons generated from the emission layer and may be formed of bis(2-methyl-8-quinolinolate)-(4-hydroxy-biphenyl)-aluminium. The hole transport layer facilitates hole transport to the emission

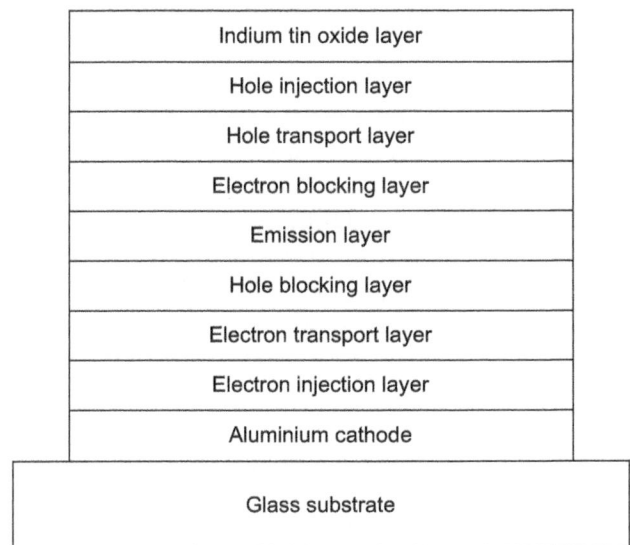

Figure 4.5 Complex structure of an OLED or PHOLED.[14]

layer and is a low molecular weight material such as N,N′-bis(naphthalene-1-yl)-N,N′-bis(phenyl)benzidine. The hole injection layer is formed by thermal evaporation of an inorganic semiconductor, for example, vanadium pentoxide or tungsten oxide. Established techniques such as spin coating, ink-jet printing or laser-induced thermal imaging can be used to deposit layers in the OLED or PHOLED.

The development of OLED displays does not rely just on fabrication of materials with tailored semiconducting properties. Of importance is the electronic circuitry, in particular, the driver circuits required to power the device. Organic light-emitting displays are described as either passive matrix organic light-emitting displays (PMOLED) or active matrix organic light-emitting displays (AMOLED), depending on the method of driving the organic light-emitting diodes. AMOLEDs contain pixels positioned at the crossing regions of data columns and scan rows. Each pixel has an OLED and pixel circuit for driving the OLED. The pixel circuit generally contains a switching transistor, a driving transistor and storage capacitor. AMOLEDs can operate with lower power consumption than LCD displays, have a thinner profile and are popular choices for portable devices. Developments in driver circuits have been described by Kwak,

Kane and Wang.[15–17] Thin film transistors (TFT) used in AMO-
LEDs are classified as either amorphous silicon (a-Si)TFT or
polycrystalline silicon (p-Si)TFT that are deposited by established
semiconductor fabrication routes.[18] Active matrix driving
schemes are constantly driven and unlike passive matrix driving
schemes they do not exhibit a 'smearing effect' when images
change quickly.[19] In general, OLEDs are capable of manufacture
by TFT-LCD processes, an important factor in the choice of these
materials for displays.[20] Examples of processes for device fabrication are given by Lee and Lee and by Johnson and
Fisekovic.[21,22]

4.5 SUMMARY

Organic light-emitting diodes have a history dating back to the
early 1950s and include small molecule OLEDs (SMOLED) and
PLEDs where electroluminescence arises within a conjugated
semiconducting polymer based on poly(p-phenylenevinylene).
Devices based on active matrix organic light-emitting diodes
(AMOLED) can be fabricated by methods used in the semiconductor industry and have low power consumption and a thin
profile, which are popular characteristics for portable devices.

REFERENCES

1. A. Bernanose, Electroluminescence of organic compounds,
   Br. J. Appl. Phys., 1955, **6**, S54–S55.
2. E. F. Gurnee and R. T. Fernandez, Organic electroluminescent phosphors, *United States Patent*, 3 172 862, 1965.
3. H. P. Kallmann and M. Pope, Current conducting device,
   United States Patent, 3 247 427, 1966.
4. D. F. Williams, Electroluminescent device with light emitting aromatic, hydrocarbon material, *United States Patent*,
   3 621 321, 1971.
5. A. J. Heeger, A. G. MacDiarmid, C. K. Chiang and H.
   Shirakawa, P-type electrically conducting doped polyacetylene film and method of preparing same, *United States
   Patent*, 4 222 903, 1980.
6. S. A. Van Slyke and C. W. Tang, Organic electroluminescent
   devices having improved power conversion efficiencies,
   United States Patent, 4 539 507, 1985.

7. J. Shi, C. H. Chen, S. A. Van Slyke and C. W. Tang, Organic electroluminescent devices with high thermal stability, *United States Patent*, 5 554 450, 1996.
8. C. W. Tang, C. H. Chen and R. Goswami, Electroluminescent device with modified thin film luminescent zone, *United States Patent*, 4 769 292, 1988.
9. C. H. Chen, C. W. Tang, J. Shi and K. P. Klubek, Green organic electroluminescent devices, *European Patent*, 1010742B, 2004.
10. R. H. Friend, J. H. Burroughes and D. D. Bradley, Electroluminescent devices, *United States Patent*, 5 247 190, 1993.
11. Osram GmbH, *Introduction to OLED technology*, Osram GmbH, 2011, www.osram.com/oled.
12. M. Hack, M.-Hao M. Lu and M. S. Weaver, Organic light emitting devices for illumination, *United States Patent*, 8 100 734, 2012.
13. M. Hack, J. J. Brown, P. Levermore and M. S. Weaver, Organic light emitting device lighting panel, *United States Patent Application*, 2011/0284899, 2011.
14. S.-B. Lee, S.-W. Noh, J.-W. Seong and M.-H. Kim, Organic light emitting diode display device and method of fabricating the same, *United States Patent*, 8 294 358, 2012.
15. W.-K. Kwak, Pixel and organic light emitting display device using the same, *United States Patent*, 8 237 634, 2012.
16. M. G. Kane, Pixel circuit for an active matrix organic light-emitting diode display, *United States Patent*, 7 956 825, 2011.
17. C. Y. Wang, Driver circuit of AMOLED with gamma correction, *United States Patent*, 7 880 692, 2011.
18. S. J. Bae and J. Y. Lee, Organic electroluminescence device and fabrication method thereof, *United States Patent*, 7 963 816, 2011.
19. German Flat Panel Display Forum, *European technology: Flat panel displays*, VDMA Verlag GmbH, 6th edn, 2008.
20. H.-H. Lee and M.-C. Shih, Method of forming an organic light-emitting display with black matrix, *United States Patent*, 7 781 348, 2010.
21. J. H. Lee and J. S. Lee, Organic light-emitting diode display device and method of fabricating the same, *United States Patent*, 8 227 983, 2012.
22. M. T. Johnson and N. Fisekovic, Oled display device, *United States Patent*, 8 294 641, 2012.

CHAPTER 5

Liquid Crystals and Liquid Crystal Displays

5.1 INTRODUCTION

Electronic displays have become an increasingly ubiquitous feature of consumer products such as pocket calculators, digital watches, digital cameras, flat panel televisions with large screens, smartphones with small displays, portable computers such as laptop and notebook computers, portable video players and in-vehicle navigation displays. Electronic displays frequently incorporate liquid crystals that exhibit an electro-optic effect, have an electrically switchable molecular arrangement and also satisfy demand for low power consumption. These compact displays replace conventional screens using bulky cathode ray tubes. Liquid crystal displays represent an apex of interdisciplinary activities embracing chemical synthesis, materials science, physics and electrical engineering. However, liquid crystals have, as was noted for light-emitting diodes, a history that dates back many decades. It was Friedrich Reinitzer who first observed a mesophase (liquid crystal phase) for a cholesterol derivative in 1888.[1,2] This derivative became cloudy and viscous on melting at 145 °C, after which it became isotropic (disordered) and clear at 179 °C. Lehmann, who was aware of Reinitzer's observations, used

Exploring Materials through Patent Information
By David Segal
© David Segal, 2015
Published by the Royal Society of Chemistry, www.rsc.org

the phrase 'fliessende krystalle' to highlight characteristics of both the liquid state and crystalline materials, hence the phrase 'liquid crystal'.[2,3] The liquid crystalline phase resides in the temperature range between the first melting point temperature and the temperature at which isotropic liquid is formed.

It is common nowadays to see advertisements for televisions with LCD screens or smartphones with LCD displays, but liquid crystals were relatively unknown to the general public and scientific communities until the 1960s. The development and uses for liquid crystalline materials in electronic displays is described here with emphasis on publications from the patent literature, a primary source of technical information.

5.2 LIQUID CRYSTALS: AN OVERVIEW

Liquid crystalline phases exist between crystalline solid phases and the fully disordered liquid state and are associated with long-range molecular ordering.[4,5] Molecules in these phases combine liquid-like properties such as low viscosity with crystal-like properties, for example, anisotropy. Liquid crystalline molecules discovered by Reinitzer are thermotropic and there are three types of thermotropic phases: (i) smectic, (ii) nematic and (iii) cholesteric. Molecular ordering in the smectic mesophase is of a lamellar kind, where molecules are arranged side-by-side in each layer and have an elongated or cigar-like shape. Ordering in the nematic mesophase occurs along the long molecule axis and these liquid crystals have a thread-like structure. The direction along which molecules are oriented is known as a director.[6] In the cholesteric mesophase, molecules are ordered in nematic-layers and the director is rotated around a helix characterised by a pitch, p of the order of the wavelength of visible light (Figure 5.1). Selective reflection of circularly polarised light makes the material look coloured.

Williams demonstrated the feasibility of using liquid crystals as electro-optical elements for display devices.[2,7] Two transparent quartz plates were spaced about 20 μm apart. Transparent indium tin oxide electrodes were attached to each plate and faced each other. The space between the two plates was filled with a thermotropic nematic compound, for example, p-azoxyanisole, anisaldazine, 4,4'-dimethoxystilbene or dibenzalbenzidine or

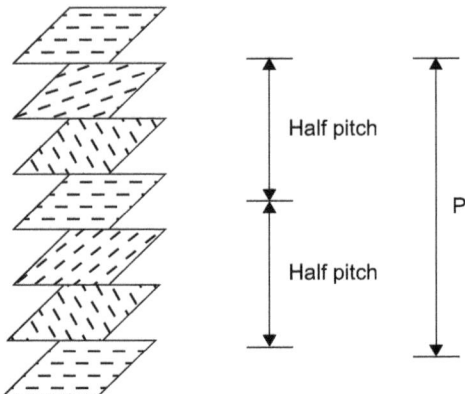

Figure 5.1 Schematic diagram for ordering in a cholesteric liquid crystal.[2]

4,4′di-n-heptoxy-azoxybenzene. The plates were maintained in the temperature range in which materials exhibited the nematic mesophase. In the absence of an applied electric field, the device was relatively transparent to light. When an electric field was applied to the liquid crystal material above a threshold value, about 1000 Vcm^{-1}, the device appears to darken in the region of the field. Williams described the darkening due to light scattering by domains containing liquid crystal molecules which align themselves in the electric field. The domains change direction above the threshold electric field but all molecules in a domain retain the same direction. Williams proposed a device for displaying patterns or information using this electro-optical effect.

At about the same time as Williams demonstrated electro-optic properties of liquid crystals, Weimer was laying the foundations for complementary metal-oxide-semiconductor (CMOS) circuits that were suitable as driving circuits for pocket calculators with liquid crystal displays, as they used low power and could run off batteries.[8]

Practical LCD devices require liquid crystals with a low transition temperature. Goldmacher and Heilmeier used a mixture of 50 weight % p-n-ethoxybenzylidene-p′-aminobenzonitrile and 50 weight % p-n-butoxybenzylidene-p′-aminobenzonitrile that had a crystal-nematic transition temperature of 41 °C and a nematic-isotropic transition temperature of 109 °C.[9] A schematic diagram for a cell that operates with cross polarisers is shown in

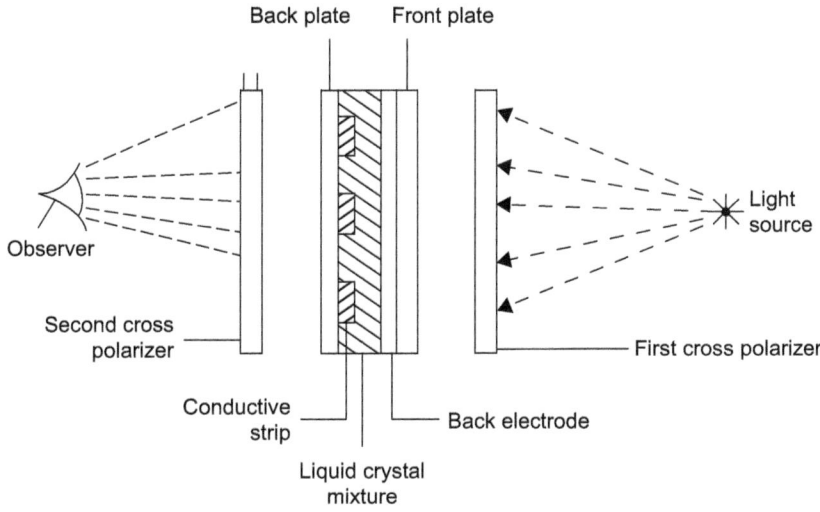

Figure 5.2   Schematic view of electro-optic cell with cross polarisers.[9]

Figure 5.2. The cell utilises birefringent properties of the nematic liquid crystal composition and rotation of the plane of polarisation for incident light. Thus, optical anisotropy is a feature of liquid crystals. In use, light is polarised after passage through the first polariser and then transmitted through the liquid crystalline material. Light passes through the second polariser so that the cell appears uniformly bright to an observer when an electric field is absent. This is because the domains are randomly oriented, causing rotation of polarised light, which then passes through the second polariser. However, when an electric field is applied across the device, liquid crystal domains align so that molecules are parallel to the direction of incident light. The polarisation plane is not rotated so that the second polariser, which is crossed relative to the first polariser reduces the transmitted light intensity. An observer sees dark regions near the applied field. Thus, the applied electric field modulates the light intensity. Pleochroic dyes such as methyl red were mixed in with the nematic material. These dyes absorb light when the direction of polarisation for linearly polarised light is in the direction of their long axis and transmit light when the direction of polarisation is not in the direction of the long axis.[2]

Table 5.1 Novel liquid crystal compositions based on $X(C_6H_4)CHN(C_6H_4)Y$ where X and Y are terminal groups.[10]

X	Y	Crystal-mesomorphic transition (°C)	Mesomorphic-isotropic liquid transition (°C)
CH_3CH_2CO	CH_3O	86	118
$CH_3(CH_2)_2COO$	CH_3O	86	119
$CH_3(CH_2)_2COO$	$C_6H_{13}O$	86	120
C_4H_9O	CH_3COO	82	113
$C_5H_{11}O$	CH_3COO	74	82
$C_6H_{13}O$	CH_3COO	88	109
$C_8H_{17}O$	CH_3COO	80	105
$C_9H_{19}O$	CH_3COO	86	100
CH_3O	$CH_3(CH_2)_2COO$	49	113

Goldmacher and Castellano identified nematic phases with low transition temperatures based on the composition $X(C_6H_4)CHN(C_6H_4)Y$, where X and Y are terminal groups.[10] Examples of these materials are listed in Table 5.1, which illustrates both the low crystal-mesomorphic transition temperature and the wide temperature range within which the mesophase is stable.

Mixtures of cholesteric and nematic liquid crystalline phases exhibited storage of the light scattering when the applied electric field was removed and the time for the mixture to return to its transparent state was extended to several weeks.[11] Return to the transparent state, described as erasure, could be reached in milliseconds when a high frequency a.c. field was applied to the liquid crystal mixture. A composition that exhibited storage consists of equal weight ratios of anisylidene-p-aminophenylacetate, p-n-butoxybenzylidene-p-aminophenylacetate and anisylidene-p-aminophenylbutyrate combined with a mixture containing 23 weight % of cholesteryl chloride and 77 weight % of cholesteryl oleate.

Heilmeier showed that the absorption spectrum of a pleochroic dye, known as a guest material could be controlled in an applied electric field when mixed with a nematic liquid crystal known as a host material.[12] For example, a mixture of methyl red and the nematic liquid crystal, butoxy benzoic acid, was held between 147–163 °C, while a small d.c. potential, about 10 V was applied across the material. This guest-host mixture varied in colour from orange to yellow, depending on the direction of polarisation for incident light with respect to the molecular axes of dye molecules. Mixtures of butoxy benzoic acid and indolphenol blue varied in colour from deep blue to pale blue. Heilmeier disclosed that the ability to switch colours at low voltage was useful for television transmission but the advent of colour television systems was some years away. Dyes and liquid crystal hosts were not stable over time in applied fields and heating was required to keep the host in a nematic phase. It was found that certain nematic liquid crystals such as anisylidene para-aminophenylacetate are transparent but scatter light and appeared milky white in an applied electric field.[13] This light scattering was termed dynamic scattering and was considered to arise from turbulence in the nematic phase brought about by ion transport in the liquid crystalline phase. A recovery or turn-off time of around 3 milliseconds made these liquid crystals attractive choices for alphanumeric displays. A selection of other patent documents associated with Heilmeier is listed in References 14–21 inclusive.[14–21]

Janning made contributions to the eventual exploitation of liquid crystal displays. Firstly, in a practical display where alphanumeric characters could be clearly illuminated and, of importance, an oblique alignment or oblique evaporation process.[22,23] Two glass plates, nearly upright, were placed in a vacuum chamber and exposed to an alignment material such as platinum vapour. A layer of nematic liquid crystals such as 4-methoxy 4′n-butyl-benzylidene-aniline (MBBA) was then introduced between the plates. MBBA molecules are deposited onto the alignment film and form an adherent coating that aligns molecules in the direction of alignment of film growth. Two alignment films on electrode coatings on glass plates are arranged by turning one plate through 90° to form a polarising cell. An alternative approach to oblique alignment was carried out by rubbing two electrodes with a cotton cloth and arranging surfaces to face one another with their rubbing directions initially perpendicular.[24] A layer of nematic liquid crystals was introduced between the plates and molecules adjacent to the surfaces align with their axes following the rubbing directions on the surface. Molecules at increasing distance from the surface undergo a 90° progressive twist in the direction of molecular axes through the layer, thus a quarter twist of a helix. When an electric field was applied across the plates, spiral twisting ceased as molecules changed alignment, blocking light transmission from the cell.

It is the optical anisotropy and change in alignment direction of molecules by an applied electric field that blocks or transmits light and allows liquid crystals to display colours and images.

A modified rubbing method was developed by Raynes to allow for interaction between molecules and glass plate, leading to improved surface coverage.[25] References in this chapter have, so far, summarised early attempts in the development of liquid crystals and their use in displays. These attempts identified three electro-optical effects: (i) domain formation, (ii) guest-host interactions and (iii) dynamic scattering. However, practical devices had to wait for advances in materials, in particular liquid crystals with appropriate electro-optical properties.

5.3 LIQUID CRYSTALS: ADVANCES IN MATERIALS

A key advance in materials development was made by Helfrich and Schadt at Hoffmann-La Roche,[26,27] when they identified a twisted nematic structure that was suitable for displays but only if the liquid crystal has a positive dielectric anisotropy, that is, the dielectric constant along the long axis of the molecule, $\varepsilon_{parallel}$ is greater than the dielectric constant in the perpendicular direction, $\varepsilon_{perpendicular}$, that is:

$$\varepsilon_{parallel} > \varepsilon_{perpendicular} \tag{5.1}$$

An example of a liquid crystal with positive dielectric anisotropy is n-4′-ethoxybenzylidene-4-aminobenzonitrile (PEBAB). As explained in Reference 27, the optical activity of liquid crystals with positive dielectric anisotropy can be controlled by an applied electric field.[27] When the latter was in the direction of the helix axis and perpendicular to the plates, molecules, except those at the adherent boundary layer, align themselves parallel to the field so that the natural helical structure untwists and optical activity ceases. The helical structure is restored when the field is switched off and their realignment when $\varepsilon_{parallel} > \varepsilon_{perpendicular}$ is associated with a dielectric torque. Liquid crystals that exhibit this behaviour are known as twisted nematics. Figure 5.3 shows

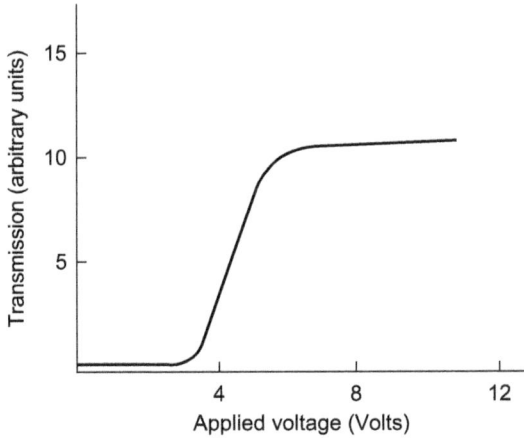

Figure 5.3 Variation of light transmission with voltage for PEBAB.[27]

the variation of transmitted light intensity with applied field for PEBAB contained as a layer with thickness 10–100 μm in a cell. A threshold of 2.5 volts corresponds to an electric field strength in the range 250–2500 V cm^{-1}. The polarisation of incident light is affected by dynamic scattering, but not by materials with positive dielectric anisotropy so that a laser light source can be used with the latter.

Fergason disclosed a liquid crystal composition with positive dielectric anisotropy that was nematic at room temperature at around the time when Helfrich and Schadt made their advance.[28] This liquid crystal composition contained 40 weight % bis-(4'-n-octyloxybenzal)-2-chlorophenylenediamine, 50 weight % p-methylbenzal-p'-n-butylaniline and 10 weight % p-cyano-benzal-p'-n-butylaniline.

Significant contributions to the field of liquid crystals were made by Gray and co-workers at the University of Hull, in particular on development of cyanobiphenyls with stable nematic phases at or near ambient temperatures.[4,5,29–32] Figure 5.4(a) shows the molecular structure of liquid crystals used in displays at the time Gray began his activities. This structure is characteristic of Schiff's bases where the A-B unit often contains a double or triple bond.[5] The presence of these bonds leads to chemical or photochemical instability, for example, susceptibility to hydrolysis by water vapour. Gray's innovative work led

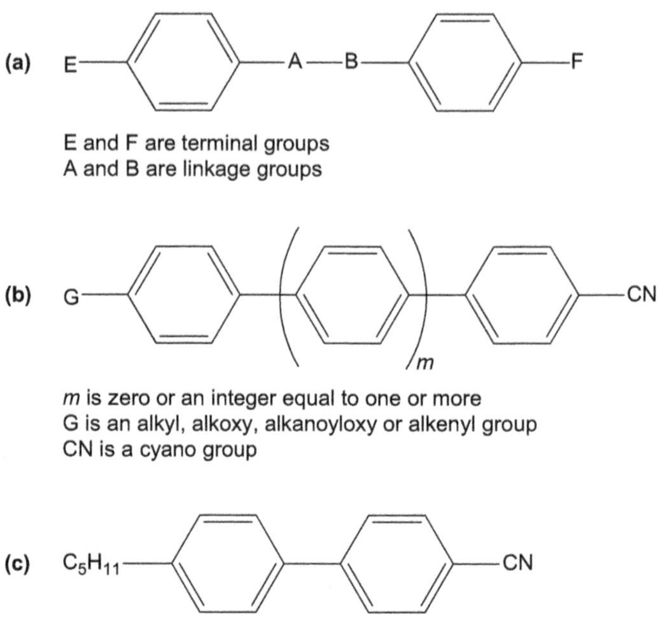

(a)   E—⟨benzene ring⟩—A—B—⟨benzene ring⟩—F

E and F are terminal groups
A and B are linkage groups

(b)   G—⟨benzene ring⟩—(⟨benzene ring⟩)$_m$—⟨benzene ring⟩—CN

m is zero or an integer equal to one or more
G is an alkyl, alkoxy, alkanoyloxy or alkenyl group
CN is a cyano group

(c) C_5H_{11}—⟨benzene ring⟩—⟨benzene ring⟩—CN

Figure 5.4 Development of cyano-biphenyl liquid crystals: (a) conventional structures, (b) molecular structures developed by Gray and Harrison and (c) representative structure.[4]

Table 5.2 Transition temperatures for liquid crystal structures based on $G(C_6H_4)(C_6H_4)CN$ where G is a terminal group.[4]

G	Crystal to nematic transition temperature (°C)	Nematic to isotropic transition temperature (°C)
C_5H_{11}	22.5	35
C_6H_{13}	13.5	27
C_7H_{15}	28.5	42
$C_5H_{11}O$	48	67.5
$C_6H_{13}O$	58	76.5
$C_7H_{15}O$	53.5	75

to incorporation of a cyano group (CN) into a biphenyl or polyphenol structure. These compounds, cyanobiphenyls (Figure 5.4(b)) had low crystalline to nematic and high nematic to isotropic transition temperatures, as shown in Table 5.2 and Table 5.3.

Twisted phase materials are widely used in displays including phase-change displays, guest-host displays, passive and active

Table 5.3 Transition temperatures for two-component mixtures based on $G(C_6H_4)(C_6H_4)CN$ where G is a terminal group.[4]

G	Mole %	Crystal to nematic transition temperature (°C)	Nematic to isotropic transition temperature (°C)
C_7H_{15}	48.5	0.5	40
C_8H_{17}	51.5		
$C_5H_{11}O$	55	21	70.5
$C_7H_{15}O$	45		
C_5H_{11}	83.5	15	42
$C_6H_{13}O$	16.5		

matrix displays, ferroelectric displays and cholesteric displays.[33] However, development of materials with improved optical properties is a buoyant area of research and development. One reason for this is that contrast is dependent on the viewing angle at which the screen is observed.[34] The pitch, p, of the helical molecule is a crucial parameter for the reflection of circular polarised light in a wavelength region dependent on the helical structure, and syntheses have concentrated on modifying this parameter particularly in the presence of chiral dopants.[35] Linsted has described a process for producing liquid crystal polymer by melt polymerisation.[36] Polymers for use in displays have been described by Ohgiri *et al.*[37] Here, a polymerisable compound was added to a liquid crystalline medium for concentrations less than 5 weight % and polymerised by, for example, exposure to ultra-violet radiation. The polymer stabilised initial alignment of liquid crystals, contributing to a wider viewing angle for the device and improved response time. Nakanishi *et al.*[38] also used a polymerisable material that aligned liquid crystals in the absence of an applied electric field. The polymer helped prevent 'image sticking', whereby a residual background pattern remained on a screen when an LCD device was driven for a long time. A further example of the detailed chemistry required in synthesis of liquid crystalline media has been given by Wittek *et al.*[39]

5.4 LIQUID CRYSTALS: DEVELOPMENTS IN DISPLAYS

Early developments in liquid crystal displays are exemplified by the work of Fergason, but at that time the availability of consumer products with flat panel displays was very limited.[40–44] Nowadays, the demand for consumer goods has resulted in continual improvements to displays. Liquid crystal devices are passive, that is, they do not emit light and require a backlight for illumination, a cold cathode fluorescent lamp (CCFL). Details of some of the recent developments in displays follow in the next section.

Kamada produced a diffusion plate for a backlight unit that did not emit annoying squeaking sounds during deformation, caused by heat from the light source.[45] Oh used a backlight unit with a light guide that prevented unwanted light transmission due to reflections off walls in a mobile device.[46] The backlight intensity was adjusted through measurement of ambient light intensity by Lowles *et al.*[47] to increase the clarity of information on a mobile display. Kuo *et al.*[48] developed a flexible colour-filter-on-array display panel. A liquid crystal layer was deposited between a solid base such as glass with thickness between 100 and 700 μm and a flexible substrate, for example, polycarbonate. After a thermal treatment the base was thinned by etching or polishing to produce the flexible unit.

Touch screen devices are becoming increasingly popular and Furuhashi and Mamba described a liquid crystal display with a capacitive touch screen.[49] As displays become larger, issues relating to the viewing angle become more significant. Park *et al.*[50] developed a multiple alignment technique for liquid crystalline molecules to improve the viewing process. In the method, an alignment film is formed on a surface onto which are deposited fluoro-polymer patterns by laser ablation. A rubbing method was used to alter the alignment directions of exposed film after which the polymer patterns were removed with a solvent. Park *et al.*[51] increased the durability of polarising plates in a humid atmosphere with prevention of light leakage and yellowing by using protective films for polyvinyl alcohol polarising plates. Detailed descriptions of liquid crystal devices and their fabrication are given in References 52–54.

5.5 SUMMARY

Liquid crystal displays are used in a wide range of consumer goods, for example, flat panel televisions, smartphones, notebook computers and in-vehicle navigation displays. Optical anisotropy and change in alignment direction of molecules in an applied electric field blocks or transmits light and allows liquid crystals to display colours and images.

REFERENCES

1. F. Reinitzer, Beitrage zur Kenntniss des Cholesterins, *Monatsh. Chem.*, 1888, **9**, 421–441.
2. H. Kawamoto, The history of liquid-crystal displays, *Proc. IEEE*, 2002, **90**(4), 460–500.
3. O. Lehmann, Über fliessende Krystalle, *Z. Phys. Chem.*, 1889, **4**, 462–472.
4. G. W. Gray and D. G. McDonnell, Cyclohexane containing derivatives of aromatic nitriles and their use as liquid crystal compounds, *United Kingdom Patent Application*, 2023136A, 1979.
5. G. W. Gray and K. J. Harrison, Substituted biphenyl and polyphenyl compounds and liquid crystal materials and devices containing them, *United Kingdom Patent Application*, 1433130, 1976.
6. J. Lub, Polymerizable liquid crystalline dioxetanes, their preparation and use, *European Patent*, 1337602B, 2006.
7. R. Williams, Electro-optical elements utilizing an organic nematic compound, *United States Patent*, 3 322 485, 1967.
8. P. K. Weimer, Insulated gate field effect devices and electrical circuits employing such devices, *United States Patent*, 3 191 061, 1965.
9. J. E. Goldmacher and G. H. Hellmeier, Nematic liquid crystal mixtures for use in a light valve, *United States Patent*, 3 499 702, 1970.
10. J. E. Goldmacher and J. A. Castellano, Electro-optical compositions and devices, *United States Patent*, 3 540 796, 1970.
11. J. E. Goldmacher and G. H. Hellmeier, Liquid crystal display element having storage, *United States Patent*, 3 703 331, 1972.
12. G. H. Hellmeier, Control of optical properties of materials with liquid crystals, *United States Patent*, 3 551 026, 1970.
13. G. H. Hellmeier, Fast self-quenching of dynamic scattering in liquid crystal devices, *United States Patent*, 3 575 493, 1971.

14. G. H. Hellmeier and L. A. Zanoni, Reduction of turn-on delay in liquid crystal cell, *United States Patent*, 3 503 673, 1970.
15. G. H. Hellmeier, Turnoff method and circuit for liquid crystal display element, *United States Patent*, 3 519 330, 1970.
16. G. H. Hellmeier and L. A. Zanoni, Electro-optical device, *United States Patent*, 3 499 112, 1970.
17. G. H. Hellmeier and J. E. Goldmacher, Liquid crystal light valve containing a mixture of nematic and cholesteric materials in which the light scattering effect is reduced when an electric field is applied, *United States Patent*, 3 650 603, 1972.
18. G. H. Hellmeier, Decreasing response time of liquid crystals, *United States Patent*, 3 575 491, 1971.
19. G. H. Hellmeier, Panel structure for matrix addressed displays, *United States Patent*, 3 603 984, 1971.
20. G. H. Hellmeier, Electro-optic device having grooves in the support plates to confine a liquid crystal by means of surface tension, *United States Patent*, 3 600 061, 1971.
21. G. H. Hellmeier, Liquid crystal display assembly having independent contrast and speed of response controls, *United States Patent*, 3 655 269, 1972.
22. J. L. Janning, Stacked plate visual display panel, *United States Patent*, 3 621 332, 1971.
23. J. L. Janning, Alignment film for a liquid crystal display cell, *United States Patent*, 3 834 792, 1974.
24. J. L. Fergason, Liquid crystal non-linear light modulators using electric and magnetic fields, *United States Patent*, 3 918 796, 1975.
25. E. P. Raynes, Liquid crystal devices, *United Kingdom Patent Application*, 1 472 247, 1977.
26. W. Helfrich and M. Schadt, Erklärung teilweisen Verzichtes, *Swiss Patent*, 532 261, 1974.
27. W. Helfrich and M. Schadt, Optical device, *United Kingdom Patent Application*, 1 372 868, 1974.
28. J. L. Fergason, Display devices utilizing liquid crystal light modulation, *United States Patent*, 3 731 986, 1973.
29. D. Coates and G. W. Gray, Biphenyl derivative compounds and liquid crystal materials and devices containing such compounds, *United States Patent*, 4 035 056, 1977.

30. G. W. Gray, S. M. Kelly, D. G. McDonnell and A. Mosley, Liquid crystal compounds and materials and devices containing them, *European Patent*, 0008188B, 1982.
31. B. M. Andrews, N. Carr, G. W. Gray and C. Hogg, Liquid crystal compounds, *European Patent*, 0097033B, 1990.
32. C. M. Waters and E. P. Raynes, Liquid crystal devices, *United Kingdom Patent Application*, 2123163A, 1984.
33. A. L. May, S. Greenfield, J. Goulding and O. L. Parri, Chiral compounds, *United States Patent*, 6 723 395, 2004.
34. M. Bremer, A. Goetz and S. Derow, Polymerizable compounds, *United States Patent*, 7 807 068, 2010.
35. F. Prechtl, S. Haremza, F. Meyer, R. Parker, V. Vill and G. Gesekus, Chiral compounds, and their use as chiral dopants for the preparation of cholesteric liquid-crystalline compositions, *United States Patent*, 6 569 355, 2003.
36. H. Clay Linstid III and V. J. Provino, Process for producing liquid crystal polymer, *United States Patent*, 6 114 492, 2000.
37. S. Ohgiri, M. Okamura, Y. Sugiyama and H. Ichinose, Liquid crystalline medium and liquid-crystal display, *United States Patent Application*, 2012/0256124, 2012.
38. Y. Nakanishi, Y. Inoue, K. Hanaoka, H. Yoshida, Y. Tasaka and K. Tashiro, Liquid crystal display device and manufacturing method thereof, *United States Patent Application*, 2012/0261846, 2012.
39. M. Wittek, B. Schuler, V. Reiffenrath, A. Manabe, E. Meyer and M. Goebel, Liquid-crystalline medium and liquid-crystal display, *United States Patent*, 8 012 370, 2011.
40. J. L. Fergason, Display devices utilising liquid crystal light modulation, *United States Patent*, 3 731 986, 1973.
41. J. L. Fergason, Liquid crystal mixtures for use in liquid-crystal displays, *United States Patent*, 3 960 749, 1976.
42. J. L. Fergason, Gasket for liquid crystal light shutters, *United States Patent*, 3 853 392, 1974.
43. J. L. Fergason and T. B. Harsch, Reflection system for liquid crystal displays, *United States Patent*, 3 881 809, 1975.
44. J. L. Fergason and D. E. Werth, Liquid crystal display assembly, *United States Patent*, 3 963 324, 1976.
45. K. Kamada, Backlight unit and liquid crystal display, *United States Patent*, 8 228 458, 2012.

46. M. R. Oh, Backlight unit and LCD having the same, *United States Patent Application*, 2012/0242928, 2012.

47. R. J. Lowles, M. A. Drader and J. Robinson, Dual-function light guide for LCD backlight, *United States Patent*, 8 279 158, 2012.

48. W.-H. Kuo, T.-C. Yang, S.-L. Lee and W.-M. Huang, Flexible liquid crystal display panel and method for manufacturing the same, *United States Patent Application*, 2012/0264243, 2012.

49. T. Furuhashi and N. Mamba, Liquid crystal display device with touch screen, *United States Patent*, 8 139 037, 2012.

50. W.-S. Park, K.-H. Uh, S.-D. Lee, J.-H. Na, Y.-T. Kim and K.-M. Koo, Methods of manufacturing alignment substrate and liquid crystal display device having the alignment substrate, *United States Patent Application*, 2011/0221098, 2011.

51. J.-H. Park, J.-S. Park, M.-S. Kim and S.-H. Kim, Polarizer and liquid crystal display using the same, *United States Patent*, 8 164 715, 2012.

52. J.-H. Jung, H.-S. Hong and B.-H. Lim, Liquid crystal display device and fabrication method thereof, *United States Patent*, 8 208 085, 2012.

53. K. Miyamoto, M. Hayashi, M. Tanahara and M. Aoki, Liquid crystal display device, *United States Patent*, 8 208 103, 2012.

54. M. Ohara, Liquid crystal display device, *United States Patent Application*, 2012/0223921, 2012.

3D Printing

6.1 INTRODUCTION

The materials described in the preceding chapters: light-emitting diodes (LEDs), quantum dots, organic light-emitting diodes (OLEDs) and liquid crystal displays (LCD) can each be characterised by a physical phenomenon. Hence, semiconductor bandgaps for LEDs, quantum confinement for quantum dots, conjugated organic polymers for OLEDs and optical anisotropy for liquid crystal displays. In addition, LEDs, quantum dots, OLEDs and LCDs share a common industrial use, namely lighting technology. Understanding these complex technical areas is difficult because they involve processes that take place at the molecular level that can be represented mathematically or schematically. The subject matter in this chapter, 3D printing, refers to manufacturing techniques and not to the effect of applied electric fields or electromagnetic radiation on materials. But a common theme unites 3D printing with LEDs, quantum dots, OLEDs and LCDs, namely, the development of 3D printing can be traced from the extensive patent literature.

6.2 MANUFACTURING TECHNIQUES

Components, metallic, ceramic and plastic, undergo a finishing step in their production before use. This step can be carried out

Exploring Materials through Patent Information
By David Segal
© David Segal, 2015
Published by the Royal Society of Chemistry, www.rsc.org

by standard machining methods such as milling, drilling and polishing so that parts meet shape, size and quality requirements. For machining, abrasives can be used, although non-abrasive methods such as laser beam machining or ion-beam machining can also be used.[1] This step removes material from the component and is known as a subtractive production or subtractive machining method.[2] Subtractive methods generate waste material and machine accuracy is reduced on tool wear, increasing costs due to the requirement for replacement tools. In addition, requirements for tooling, such as moulds and dies, for example, injection moulding dies, to bring production to market can increase the manufacturing costs due to the time required to develop the tooling.[3] 3D printing, also known as additive manufacturing (AM), is a process in which three-dimensional objects are built up layer-by-layer without the drawback of subtractive machining processes. Additive manufacturing is sometimes referred to as a rapid prototyping technique.

6.3   3D PRINTING: THE BEGINNING

Swanson and Kremer arranged for two laser beams to intersect at specific co-ordinates in a volume of polymerisable liquid.[4] Regions where the beams intersected were curable and solidified. A solution of photo-crosslinking agent such as 2-methyl-anthraquinone in a monomer, for example, vinyl acetate containing a photo-polymerisation catalyst whose absorption spectrum is similar to the crosslinking agent, was frozen and exposed to two laser beams. The monomer was polymerised and crosslinked at the point where the beams intersected. After exposure, the glass is melted and exposed solid areas are separated from the matrix. Potential difficulties with this method were reduced light intensity and poorer resolution of focused spots as the intersection point penetrated the liquid. Hull combined lithographic techniques with simultaneous execution of computer-aided design (CAD) from computer instructions to produce three-dimensional objects.[5] He described this process as stereolithography, where no tooling is required, and the process is shown schematically in Figure 6.1. A focused spot of ultra-violet light scans the top surface of photo-polymerisable liquid in a

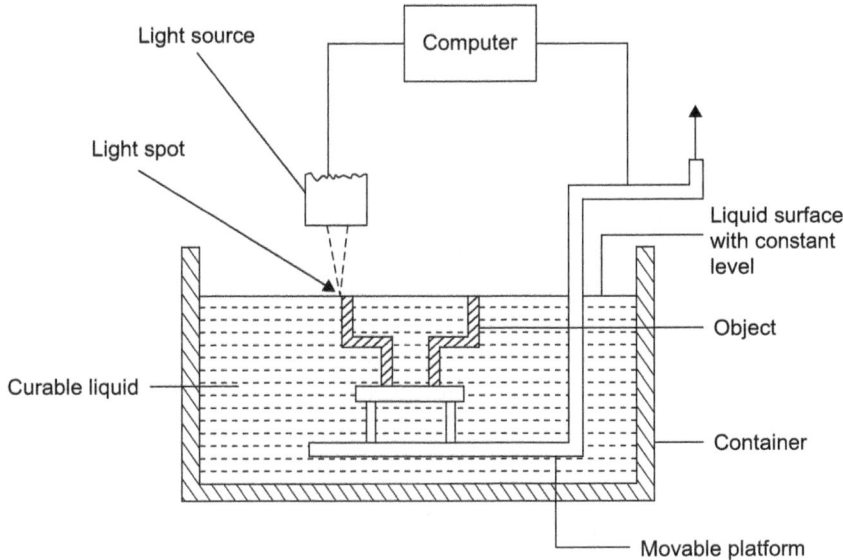

Figure 6.1 Schematic diagram for the stereolithographic process.[5]

container. The beam is moved across the surface in a predetermined pattern determined by the CAD system. The liquid is cured where the focused beam meets the surface, producing a solid plastic layer. The latter is lowered into the container and the polymerisation process is repeated. Superposition of successive adjacent cured layers onto each other is carried out automatically as they are formed to build up three-dimensional objects.

Fused Deposition Modelling (FDM), an additive manufacturing technique, was developed by Crump, in which three-dimensional prototypes or production parts are built up layer-by-layer from a thermoplastic material using a digital representation of the component.[6,7] In stereolithography, liquid is cured by a focused beam of ultra-violet light, but in FDM a thermoplastic material in the form of a rod or reel is fed to a dispensing head, melted and dispensed through an extrusion tip at a controlled rate onto a substrate. Layers with thicknesses in the range 2.5–3000 μm are built up from molten plastic droplets that adhere to preceding layers. The position of the dispensing head is computer-controlled and the head is moved in incremental steps in a vertical direction, that is, perpendicular to the substrate to build up the three-dimensional part. The substrate is then removed to leave a free-standing component.

Selective Laser Sintering (SLS) was developed by Deckard, in which a layer of metal, ceramic or polymer powder is deposited under computer-control and then sintered by scanning a laser across the surface.[8] Another powder layer was then applied to build up a three-dimensional part. However, the term 3D printing was first used by Sachs *et al.*[9,10] In their technique, a metal, ceramic or polymer powder was deposited under computer-control to a layer of thickness between 100 and 200 μm, after which a liquid binder was supplied to the powder layer using an ink-jet printer, according to the computer model for the three-dimensional part being formed. Unbound powder is removed, after which the three-dimensional component is built up. The binder ensures that all regions of each layer used to fabricate the component are uniformly bound together.

Methods described in this section highlight the early evolution of 3D printing.[11] The general approach taken in these activities consists of slicing a three-dimensional computer model generated from a CAD system into thin cross-sections, translating the result into two-dimensional position data and using the data to control equipment for manufacture of a three-dimensional part in a layerwise manner.

6.4 APPLICATIONS OF 3D PRINTING

There is evidence from patent filing dates that 3D printing is a growing activity with over 3000 patent families filed over the past twenty years.[12] Areas of activity include (i) dental implants, (ii) hearing aids, (iii) polymer materials and curable resins, (iv) microfluidic devices, (v) printers, (vi) ceramic components and (vii) medical implants. Examples from this wide range of activities using 3D printing are illustrated below.

Jandeska Jr and Hetzner used a rapid prototype process similar to the method developed by Sachs *et al.*[9,10,13] in order to fabricate aluminium or magnesium components. Aluminium alloy powders, composition 0.4–0.8 weight % Si; <0.7 weight % Fe; balance Al, with mean particle size 80 μm were coated with copper using a gas phase precursor, copper acetylacetonate.

The powder was mixed with 4 weight % of a sintering aid with composition 50 weight % Mg; 50 weight % Al and spread as a layer onto which an organic binder was deposited. Metal powder and organic binder were deposited as alternate layers to build up a prototype part. The latter was sintered at 595 °C, a temperature at which copper reacts with the sintering aid, forming a low melting point liquid promoting liquid phase sintering of aluminium powder. An alloy composition for additive manufacturing relates to the stainless steel powders used in Direct Metal Laser Sintering (DMLS), a rapid prototyping and tooling process in which net shape parts are made in a single process.[14] A composition was developed so that a component made by DMLS had the same properties as a steel component made by conventional manufacturing methods, in particular, it underwent precipitation hardening. Complex parts were produced from 3D-CAD models by layer-wise solidification of metal powder layers. The powder composition was 0.07 weight % C; 14.0–15.5 weight % Cr; 3.5–5.0 weight % Ni; 3.0–4.5 weight % Cu. The remaining component is Fe and the median particle size was between 20 and 100 µm. A focused beam from a 200W laser produced a spot size around 120 µm and the beam was scanned across the surface to sinter metal powder, as shown schematically in Figure 6.2.

Additive manufacturing has found application for the production of aerospace components.[15,16] Metal leading edge protective strips are used to protect composite airfoils from impact and erosion damage. Their conventional fabrication involves hot forming that has multiple steps of chemical milling or machining. A high-temperature additive manufacturing process operating near 3000 °C, for example, ultrasonic welding or laser cladding, was used to deposit titanium or titanium alloy to form the near net shape protective strip on a mandrel. Heat transfer away from the mandrel prevented formation of metallic deposits with the mandrel and subsequent contamination. Fused Deposition Modelling was used to produce a z-axis test coupon containing tensile specimens with a z-axis orientation suitable for mechanical testing.[17] The latter was carried out because z-axis orientation was considered weaker than for other orientations when parts were made by additive manufacturing techniques.

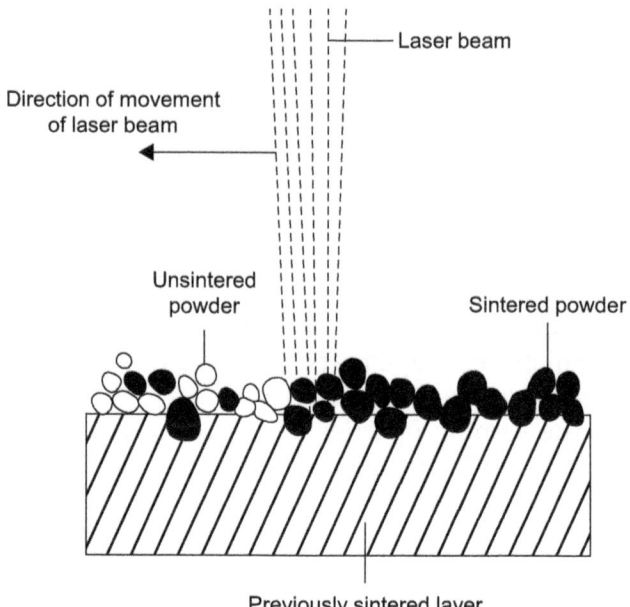

Figure 6.2 Schematic diagram of the direct metal laser sintering process.[14]

3D printing is attractive for the manufacture of aircraft structures with inter-connecting sections, such as inter-connected panels or structures that contain removable doors as the amount of adhesive required for bonding components can be reduced.[18] Selective Laser Sintering (SLS), in which powder particles distributed on a surface are sintered by a laser was used to build up the component. In a further example, James and Kulkarni fabricated a turbine blade squealer tip by Direct Metal Laser Sintering using FeCrAl alloys or oxide dispersion strengthened alloys.[19] Figure 6.3 shows a flow chart for use of direct metal laser fusion to rebuild a turbine component.[20] A design model for the component is produced on a CAD system and represented by 2D cross-sectional slices. In this rapid prototyping process, the part is directly produced by precision melting and solidification of powdered metal into successive layers of larger structures, each layer corresponding to a cross-sectional layer of the three-dimensional component.

Additive manufacturing has extended its applications to dental and medical areas. For example, a supporting structure

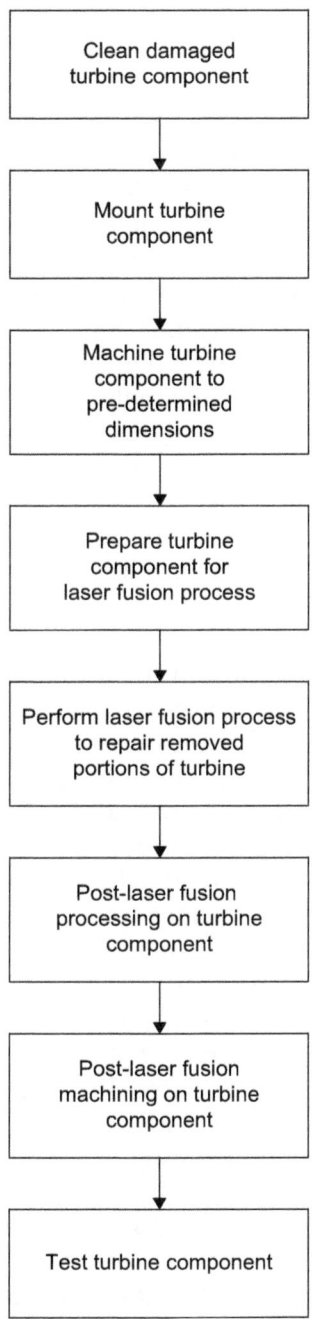

Figure 6.3 Schematic diagram showing the flowchart for repair of a turbine by laser fusion.[20]

for a dental prosthesis has been prepared by Mercelis.[21] The supporting structure is digitally designed using CAD software and is either screwed down on implants in a person's jaw or cemented to existing teeth in the jaw. Selective Laser Powder Processing is used to build up the structure, in particular Selective Laser Melting (SLM).[22] Here, the desired component geometry is built up layer-by-layer by fully melting metal powder particles, titanium or titanium alloy, with a laser (*e.g.* 100W Nd-YAG) which had a focused beam spot diameter around 100–200 μm so that new material attaches to the previous layer without use of glue or binder. Methods for obtaining a digital model for dental parts are described by Beeby *et al.*[23] while making a prosthesis which had a framework, veneer, crowns, bridges, inlays or outlays by combination of 3D printing with CAD computer-aided manufacturing was described by Wiest *et al.*[24] An exciting development of 3D printing has been the fabrication of a titanium lower jaw implant for a patient with progressive osteomyelitis in the lower jawbone.[25,26] Selective Laser Melting was used to construct the metal implant layer-by-layer.

A tissue engineering scaffold for implanting into a patient in order to promote tissue growth has also been assembled by additive manufacturing.[27] Continuous Digital Light Processing (cDLP) was used to make the implant by curing a photopolymerisable resin; pores in the scaffold had diameter in the range 50–1600 μm. Surgical guides are used in procedures in which they support bone or teeth where areas of the human body are rigid, but soft tissue guides can be used in areas of soft tissue.[28] These guides can be locked in place by screws or pins connecting the guide through a layer of soft tissue with the underlying bone. A guide frame that does not use screws or pins has been suggested by Vancraen *et al.*[28] in which additive manufacturing is combined with three-dimensional medical images of the patient. Hearing aids and customised earpieces have attracted attention from 3D printing.[29,30] The earpieces, which include housings for hearing aids, loud speakers, earplugs and headsets need to match the contours of the user's inner ear for comfort. A three-dimensional image of the inner ear can be generated by scanning, stored digitally on a computer and used in conjunction with rapid prototyping to produce customised earpieces.

The wide range of products that may be made by 3D printing include, as noted in the patent literature, (i) chocolate confectionery,[31] (ii) textile fabrics,[32] (iii) lasts for shoes,[33] (iv) models for oil well drilling equipment[34] and (v) catalyst support structures.[35]

Parts and components made by 3D printing have been described so far in this chapter. However, developments are being made continually to all aspects of the printing system. An important feature of a rapid prototyping system such as 3D printing is a shutter that opens and closes to release powder for building up layers. Between 10^3–10^5 repeated movements of the shutter may be required to fabricate a component. Van der Werff has described a shutter mechanism for an additive manufacturing system.[36] Also, Felstead and Sewell developed a powder hopper combined with a wiper blade to apply material to a substrate.[37] A three-dimensional printer with ink-jet printheads has been described, with emphasis on a powder feeder and metering to deliver powder in a controlled manner.[38] Developments in build materials that result in components with high stiffness include use of curable oligomeric species.[39] An indication of complex compositions used for build materials is shown in Table 6.1. Molten melt pools that are used to deposit material onto a substrate in rapid prototyping can result in metallic microstructures with defects such as residual stress and microcracks.[40] An alternative approach to melt pools used frictional heat to raise the temperature of a metal rod, resulting in material deposition. Although stereolithography was a key development

Table 6.1 Composition of build material for three-dimensional printing system.[39]

Component	Amount (weight %)
Oligomeric curable material	40.93
Reactive component	5.90
Diluent	40.96
Non-reactive component	5.96
Additive for curing/tackiness	1.83
Photoinitiator	4.29
Sensitizer	0.10
Inhibitor	0.02
Total	99.99

in the advance of 3D printing, developments continue to be introduced for this method. Thus, use of solid-state lasers for optimising the curing process and production, simultaneously, of two three-dimensional objects in separate chambers without intervention of an operator thus improving the process efficiency.[41,42] In rapid prototyping, the surface geometry of a three-dimensional object is represented by digital information stored in a computer, whereby the object to be fabricated is transformed into thin horizontal cross-sections that are used to construct the object. An example of software developments in ensuring that digital information gives an accurate representation of the component has been considered by Sidi and Mori.[43]

6.5 SUMMARY

3D printing, an additive manufacturing technology, has origins dating back nearly forty years. Patent filing rates indicate that 3D printing is an area of growing activity. The technology has wide applicability, that is, for aerospace structures, medical and dental implants, hearing aids and earpieces, confectionery, textile fabrics, lasts for shoes, catalyst supports and moulds for drilling equipment.

REFERENCES

1. K. Subramanian, Finishing, in *Materials Science and Technology: A comprehensive treatment*, ed. R. W. Cahn, P. Haasen and E. J. Kramer, VCH Publishers Inc., 1996, vol. 17B, Processing of Ceramics, part II, ed. R. J. Brook, pp. 215–259.
2. C. Andersson, Innovations in rapid prototyping and additive manufacturing, *European Medical Device Technology*, 2012.
3. L. J. Koppens, Injection molding, in *Concise encyclopedia of advanced ceramic materials*, ed. R. J. Brook, Pergamon Press, 1991, pp. 233–235.
4. W. K. Swanson and S. D. Kremer, Three dimensional systems, *United States Patent*, 4 078 229, 1978.
5. C. W. Hull, Apparatus for production of three-dimensional objects by stereo-lithography, *United States Patent*, 4 575 330, 1986.

6. S. S. Crump, Apparatus and method for creating three-dimensional objects, *United States Patent*, 5 121 329, 1992.
7. S. S. Crump, Modeling apparatus for three-dimensional objects, *United States Patent*, 5 340 433, 1994.
8. C. R. Deckard, Method and apparatus for producing parts by selective sintering, United States Patent, 4 863 538, 1989.
9. E. M. Sachs, J. S. Haggerty, M. J. Cima and P. A. Williams, Three-dimensional printing techniques, *United States Patent*, 5 204 055, 1993.
10. E. Sachs, A. Curodeau, T. Fan, J. F. Bredt, M. Cima and D. Brancazio, Three dimensional printing system, *United States Patent*, 5 807 437, 1998.
11. A. Levy, K. Regev, C. Rottman and S. Hirsch, System and method for additive manufacturing of an object, *International Patent Application*, WO2012/070053, 2012.
12. J. Maguire and D. Segal, 3D or not 3D. IP and 3D printing, *The TCT Magazine*, 2011, **19**(5), 37–38.
13. W. F. Jandeska, Jr and J. E. Hetzner, Aluminium/magnesium 3D printing rapid phototyping, *United States Patent*, 7 141 207, 2006.
14. T. Syvanen, O. Nyrhila and J. Kotila, Metal powder for use in an additive method for the production of three-dimensional objects and method using such metal powder, *United States Patent Application*, 2009/0047165, 2009.
15. M. W. Peretti and T. Trapp, Methods for making near net shape airfoil leading edge protection, 8 240 046, 2012.
16. M. W. Peretti and T. Trapp, Composite airfoils having leading edge protection made using high temperature additive manufacturing methods, *United States Patent Application*, 2011/0097213, 2011.
17. D. M. Deitrich and M. W. Hayes, Z-axis test coupon structure and method for additive manufacturing process, *European Patent Application*, 2420815A, 2012.
18. J. H. Wood, Methods and systems for providing direct manufactured inter-connecting assemblies, *United States Patent*, 7 977 600, 2011.
19. A. W. James and A. A. Kulkarni. Turbine blade squealer tip, *United States Patent Application*, 2012/0034101, 2012.

20. D. Mittendorf, D. Ryan, D. G. Godfrey, M. C. Morris and H. Kington, Methods for repairing turbine components, *United States Patent Application*, 2012/0222306, 2012.
21. P. Mercelis, Supporting structure for a prosthesis, *International Patent Application*, WO2010/139031, 2010.
22. J.-P. Kruth, I. Naert and B. Vandenbrouche, Procedure for design and production of implant-based frameworks for complex dental prostheses, *United States Patent Application*, 2008/0206710, 2008.
23. D. Beeby, I. Ainsworth, R. K. Revanur and G. D. Rayner, Digitization of dental parts, *United States Patent Application*, 2012/0041740, 2012.
24. T. Wiest, C. Weiss, S. Dierkes and H. Laschutza, Method and system for producing a dental prosthesis, *United States Patent Application*, 2009/0026643, 2009.
25. P. Mercelis, J. Van Vaerenbergh and W. Van de Perre, Method for manufacturing thin-walled structures in layers, *International Patent Application*, WO2012/103603, 2012.
26. K. Roberts, Layerwise builds the world's first patient-specific lower jaw implant with AM, *The TCT Magazine*, 2012.
27. D. H. Dean, J. E. Wallace, A. G. Mikos, M. Wang, A. Siblani, K. Kim and J. P. Fisher, Continuous digital light processing additive manufacturing of implants, *International Patent Application*, WO2012/024675, 2012.
28. W. Vancraen, L. J. Keppler, B. Geekelen and J. Dille, Surgical, therapeutic or diagnostic tool, *United States Patent Application*, 2012/0234329, 2012.
29. S. Parsi, Method of assembling a hearing aid, *United States Patent Application*, 2011/0289764, 2011.
30. N. Deichmann, T. Clausen, R. Fisker and C. V. Barthe, Method for modelling customized earpieces, *United States Patent*, 8 032 337, 2011.
31. A. Zimmerman, D. F. Walczyk, S. S. Crump and J. S. Batchelder, Additive manufacturing system and method for printing customized chocolate confections, *United States Patent Application*, 2012/0251688, 2012.
32. J. Kyttanen and J. Evenhuis, Method and device for manufacturing fabric material, *International Patent Application*, WO03/082550, 2003.

33. S. Dulio and E. Ratti, Production procedure for lasts for the manufacture of shoes, *International Patent Application*, WO2009/000371, 2008.

34. B. D. Calnan and V. R. C. M. J. Sillen, Molds, downhole tools and methods of forming, *United States Patent*, 7 832 457, 2010.

35. R. D. Coupland, Catalyst manufacturing method, *International Patent Application*, WO2012/032325, 2012.

36. J. J. van der Werff, Additive manufacturing system, shutter and method of building a product, *European Patent Application*, 2458434A, 2012.

37. M. W. Felstead and N. T. Sewell, Manufacturing device and method, *International Patent Application*, WO2010/061174, 2010.

38. T. Davidson, R. A. Phillips and D. B. Russell, Three-dimensional printer, *United States Patent*, 8 017 055, 2011.

39. P. Xu and J. Stockwell, Build material and applications thereof, *United States Patent Application*, 2012/0231232, 2012.

40. W. V. Twelves, Jr, W. Lin and D. G. Alexander, Solid state additive manufacturing system, *United States Patent Application*, 2009/0200275, 2009.

41. G. Cooper, Stereolithography systems and methods using internal laser modulation, *International Patent Application*, WO2012/074986, 2012.

42. B. Wahlstrom and D. F. Hunter, Rapid prototyping and manufacturing system and method, *United States Patent*, 7 585 450, 2009.

43. I. Sidi and G. Mori, Method and system enabling 3D printing of three-dimensional object models, *International Patent Application*, WO2011/042899, 2011.

CHAPTER 7

Healthcare

7.1 INTRODUCTION

In recent years, complex and expensive equipment as well as techniques and tests have been introduced into healthcare systems such as the National Health Service in the United Kingdom in order to aid medical diagnoses. For example, magnetic resonance imaging (MRI), computerised tomography, ultrasound and remote patient monitoring. These innovations often involve some aspects of materials and their properties, for example, magnets for MRI machines, transducers for ultrasound machines. However, it may not be appreciated by patients awaiting treatment that these innovations require input from diverse disciplines including chemistry, physics, materials science, mathematics and engineering. These disciplines underpin features of inventions generally that are described in the worldwide patent literature. 3D printing of a metallic implant for the lower jaw was described in Chapter 6. Additive manufacturing falls within manufacturing and engineering activities, although its application as illustrated here now extends to medical devices and hence healthcare. In this chapter, an overview is presented of selected developments in medical technology based, when available, on the patent literature.

Exploring Materials through Patent Information
By David Segal
© David Segal, 2015
Published by the Royal Society of Chemistry, www.rsc.org

7.2 BLOOD GLUCOSE MONITORING

7.2.1 From Tablets to Photometers

Diabetes is a major healthcare problem for governments in advanced economies, due to the financial costs for healthcare systems. It has been estimated that 366 million people worldwide had this condition in 2011 and that this figure will grow to 550 million by 2030, with associated annual costs of US$ 500 billion.[1] In addition, approximately 2.8 million people in the United Kingdom had the condition in 2010, a two-fold increase compared to figures for 1996.[2] In 1921 Banting and co-workers identified insulin, the pancreatic hormone that is deficient in diabetes. Nowadays, self-monitoring with blood glucose meters to determine whether insulin doses are required is accepted among the general public.

Tests for qualitative determination of glucose in urine were developed by Trommer (1841) and by Von Fehling in 1848.[2] The tests relied on reducing properties of glucose with alkaline cupric sulphate to produce coloured cuprous oxide. Although the method has been improved over the years, its disadvantage was that an external heat source was required to develop the colour. Compton and Treneer introduced a self-testing kit for glucose in urine that used a tablet containing four components: (i) sodium hydroxide, (ii) citric acid, (iii) sodium carbonate, and (iv) copper sulphate.[3] When the tablet was added to the urine sample, an increase in temperature caused the solution to boil. Glucose in urine was oxidised while Cu(II) ions were reduced, producing a change in colour from blue through green to yellow and finally to orange and the colour was then compared to a standard colour chart. Hence, the tablet had a built-in heat supply.

Free made a significant advance in testing for glucose in urine.[4] A combined mixture was formed from: (i) an enzyme, glucose oxidase, that can convert glucose to gluconic acid, also forming hydrogen peroxide, H_2O_2; (ii) the enzyme peroxidase that oxidises certain dyes in the presence of H_2O_2 causing a colour change; (iii) a dye or indicator, for example, orthotolidine dihydrochloride; and (iv) gelatin. The mixture was applied to a substrate, for example, paper strips, and dried. Strips turned blue in less than a minute when immersed in a solution containing glucose. Further development led to a dry strip

containing a semi-permeable membrane such as cellulose acetate for determining glucose concentrations in blood.[5,6] The
membrane allowed glucose molecules in a drop of blood to pass
through but red blood cells were trapped on the membrane and
then washed away. The colour in the strip was compared to a
standard colour chart. Clemens developed an automatic and
portable reflectance meter for measuring reflected light intensity
from the surface of solid strips using a photoelectric cell; results
were presented on an analogue scale.[7] Reactions that take place
on the dry strip can be summarised as follows:

$$\text{Glucose} + O_2 \xrightarrow{\text{Glucose oxidase}} \text{Gluconic acid} + H_2O_2 \qquad (7.1)$$

$$H_2O_2 + \text{oxidisable dye} \xrightarrow[\text{activity}]{\text{Enzyme with peroxidase}} \text{oxidised dye} + H_2O$$
$$(7.2)$$

Photometric methods for determination of blood glucose concentrations are described in more detail by Clarke and Foster.[2]

7.2.2 Electrochemical Sensors

Electrochemical methods for determining sugar concentrations
in blood are an alternative approach to photometric techniques.
The principle is explained by the following example.[8,9] An electrode is formed by attaching a carbon foil onto a glass strip, after
which an electron-transfer mediator, 1-1'-dimethylferrocene is
deposited from a toluene solution onto the carbon electrode.
Then an enzyme, glucose oxidase, is bonded to the surface of the
ferrocene derivative. The reference silver electrode is coated with
silver chloride and the electrode structure is illustrated in
Figure 7.1. The carbon electrode can be protected from tissue
fluid components by a membrane such as polycarbonate while
the mediator replaces oxygen in the glucose oxidase reaction
(Equation 7.1). This enzyme oxidises glucose, resulting in electron transfer from glucose to the mediator, thus reducing the
ferrocene derivative. The reduced mediator is then reoxidised at
the electrode to generate a current that can be correlated to the

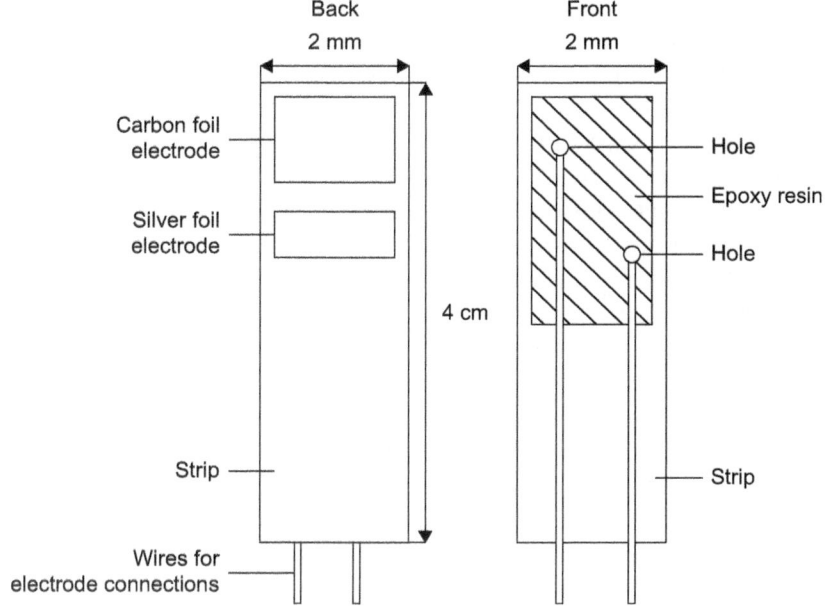

Figure 7.1 Schematic diagram of an electrochemical sensor.[9]

glucose concentration. The electrochemical sensor could be incorporated inside a holder resembling a pen.

A flame-hydrolysed silica powder that had been treated with methyl dichlorosilane to impart both hydrophobic and hydrophilic regions on the particle surface was dispersed in a solution of hydroxyethylcellulose.[10,11] The mediator, potassium hexocyanoferrate and glucose oxidase were added to the mixture that was then printed onto a carbon electrode, supported on a polyester substrate. The balance of hydrophobicity and hydrophilicity in the silica filter produced a sensor whose response was dependent on the diffusion rate of glucose molecules and not on the rate at which the enzyme oxidised glucose; performance for the disposable strip was temperature independent. Established methods for semiconductor fabrication, particularly photolithography, have been used to produce sensors for measuring blood glucose concentrations.[12] Glucose oxidase was used in conjunction with an osmium-based mediator, $(Os(\textsc{ii})(bpy)_2(im)Cl)Cl$ where im refers to imidazole and bpy refers to 2-2'-bipyridine.

So far in this chapter, methods for determining blood glucose concentrations have been described but other analytes in blood are also important. Hence, rapid determination of glycosylated proteins in blood is important for the treatment of diabetes.[13] An amperometric glycosensor was developed that measured the concentration of fructosamine derivatives formed from non-enzymatic glycosylation of proteins in blood. Two mediators were used. Representative examples of the first mediator are dimethylferrocene, tetrathiafulvalene and N-methylphenazinium, while the second mediator that was reduced by fructosamine derivates was chosen from nitroblue tetrazolium, methylene blue and thionin or mixtures of them. An overview of the technology behind glucose meters is described by Hönes *et al.*[14]

Recent applications for electrochemical sensors include a non-invasive method for analysing glucose in exhaled breath by a portable monitoring device containing a glucose sensor and storage of encoded data such as calibration data on biosensor test strips.[15,16]

7.3   POINT-OF-CARE MEASUREMENTS

Tests for multiple analytes are performed on biological samples for diagnosis, screening, forensic analysis and drug testing.[17] Laboratory testing is expensive as trained technical staff are required and there is a demand for point-of-care blood analyses at scenes of accidents, in intensive care and in surgeries. Lauks and co-workers produced a hand-held analyser with a disposable cartridge, known as an I-STAT® analyser to measure blood chemistry using a few drops of blood.[18-20] Analytes measured include glucose, urea, electrolytes and blood gases, namely oxygen and carbon dioxide. This analyser is characteristic of microfluidic devices, where fluid is moved through narrow channels continuously over a series of microfabricated sensors. The analysis is automatic and results are displayed on a screen.[21]

7.4   REMOTE PATIENT MONITORING

Rising healthcare costs are of concern to hospital administrators, governments and taxpayers. One reason for increased costs arises from labour charges for highly trained staff.

Advances in wireless, networking and electronics technologies are impacting on medical technology, particularly on measurement of physiological data, namely, vital signs and these advances have the potential to limit costs for medical treatment. Vital signs measurements include blood pressure (systolic, diastolic), blood oxygen saturation levels (SpO_2), heart rate, temperature and respiratory rate. Physiological data are conventionally obtained in a clinical environment, but there are potential cost benefits for self-testing, as well as medical benefits if measurements are recorded continuously, for example, when monitoring an athlete's performance. Russell *et al.*[22] made a monitoring device incorporating a housing with sensor pads that could be attached to a shirt for measuring vital signs, namely heart rate, breathing rate and temperature. Measurements were stored in computer memory and processed by an algorithm or wirelessly transmitted to a central computer for analysis. Banet *et al.*[23] produced a vital signs monitor that could be worn on the wrist and readings were taken continuously. Results were displayed on the device but could be wirelessly transmitted to a hospital's information technology network.

A non-contact method for measuring physiological parameters used a Laser Doppler Vibrometer (LDV).[24] A low power (1mW) red laser beam was directed onto a skin region and reflected light passed into the vibrometer, where interference took place with a reference beam. Movement of the skin surface modulates reflected laser light by means of a Doppler shift in wavelength compared to the original laser beam. Vital signs could be deduced from velocities associated with this movement. In a typical oximeter, two light beams, one red and centred around 600 nm, the other infra-red and centred at 940 nm are transmitted through a patient's skin. The transmitted light intensity is different for each wavelength. Oxygenated blood tends to absorb infra-red, whereas deoxygenated blood absorbs red light. The absorption of infra-red light relative to red light increases with oxygenated blood levels and the ratio of absorption coefficients could be used to determine oxygen saturation of the blood. Hull configured a wireless pulse oximeter where data from a sensor could be displayed on a Web page.[25] Blood pressure was measured continuously and non-invasively using a cuffless monitor, namely a sensor panel attached to a patient's head by recording the reflected light intensity from blood vessels.[26] A processor worn by the patient analysed the blood pressure readings that could be transmitted wirelessly to a web page. Alerts were sounded as an alarm when the blood pressure was above or below pre-determined set values.

Geva *et al.* developed a monitoring system in which data from a heart monitor attached to the patient are sent to a remote monitoring centre in a hospital or call centre or doctor's surgery over the mobile phone network.[27] Results are analysed and when physiological conditions are detected they are transmitted to medical professionals over the mobile phone network. Small sensor patches for heart monitoring are described by Parton.[28]

A remote patient monitoring system is shown schematically in Figure 7.2. In this example, measurements of blood glucose concentration are made by a patient at home. Results are transmitted to a patient terminal and then forwarded to a remote server that can be accessed by medical professionals either from an office or remotely. Information can then be sent back to the patient terminal. Proprietary equipment that acts as a patient terminal is described in references 30 and 31. Remote patient monitoring also covers detection of falls at home by elderly residents.[32] Sensors detect movements and if no movement was detected in a set time interval due to a fall then an alarm signal

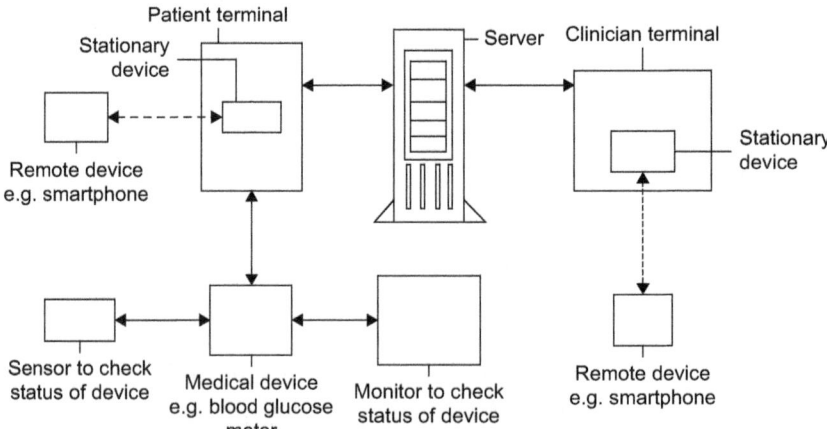

Figure 7.2   Schematic diagram of a remote potential monitoring system.[29]

was transmitted to a central control unit. Mobile phone technology with hand-held devices is making inroads into many aspects of everyday life and also inroads into medical technology.[33-35] Exchange of medical information between portable devices and computer servers is an extension to exchange of information between servers and desktop computers.

7.5 SUMMARY

Advances in wireless networking and electronics technologies are impacting on medical technology, particularly on measurement of physiological data. In addition, advances in materials have resulted in devices for routine self-testing of blood glucose concentrations. Publicly accessible patent literature sheds light on the development of healthcare technology.

REFERENCES

1. W. Betten, The state of diabetes device technology and intellectual property, *European Medical Device Technology*, 2012, 20–24.
2. S. F. Clarke and J. R. Foster, A history of blood glucose meters and their role in self-monitoring of diabetes mellitus, *Br. J. Biomed. Sci.*, 2012, **69**(2), 83–93.
3. W. A. Compton and J. M. Treneer, Tablet and method of dissolving same, *United States Patent*, 2 387 244, 1945.
4. A. H. Free, Composition of matter, *United States Patent*, 2 848 308, 1958.
5. A. H. Free, Method for determining glucose in blood, *United States Patent*, 3 061 523, 1962.
6. E. C. Adams, Jr and R. R. Smeby, Diagnostic test device for blood sugar, *United States Patent*, 3 092 465, 1963.
7. A. H. Clemens, Reflectance meter, *United States Patent*, 3 604 815, 1971.
8. H. A. O. Hill, I. J. Higgins, J. M. McCann, G. Davis, B. L. Treidl, N. N. Birket, E. V. Plotkin and R. Zwanziger, Printed electrodes, *European Patent Application*, 0351891A, 1990.
9. I. J. Higgins, J. M. McCann, G. Davis, H. A. O. Hill, R. Zwanziger, B. L. Treidl, N. N. Birket and E. V. Plotkin, Sensor electrode systems, *European Patent*, 0127958B, 1992.

10. J. F. McAleer, D. Scott, G. Hall, M. Alvarez-Icaza and E. V. Plotkin, Disposable glucose test strip and method and compositions for making same, *United States Patent*, 5 951 836, 1999.
11. J. F. McAleer, D. Scott, G. Hall, M. Alvarez-Icaza and E. V. Plotkin, Disposable glucose test strips and methods and compositions for making same, *United States Patent*, 5,708,247, 1998.
12. E. R. Diebold, R. J. Kordal, N. A. Surridge and C. D. Wilsey, Electrochemical sensor, *United States Patent*, 5 437 999, 1995.
13. P. Shieh, Determination of glycoprotein and glycosylated hemoglobin in blood, *United States Patent*, 6 054 039, 2000.
14. J. Hönes, P. Müller and N. Surridge, The technology behind glucose meters: Test strips, *Diabetes Technol. Ther.*, 2008, **10**(1), S10–S26.
15. R. J. Melker, D. G. Bjoraker, D. M. Dennis, J. D. Stewart, C. D. Batich, M. M. Booth, J. F. Horn, Jr and R. E. Youngblood, Condensate glucose analyses, *United States Patent*, 7 914 460, 2011.
16. H. Groll, M. J. Celentano and S. K. Moore, System and method for coding information on a biosensor test strip, *United States Patent*, 8 206 565, 2012.
17. C. J. Miller and J. L. E. Campbell, Method for measuring an analyte in blood, *United States Patent*, 8 168 439, 2012.
18. I. R. Lauks, H. J. Wieck, M. P. Zelin and P. Blyskal, Disposable sensing device for real time fluid analysis, *United States Patent*, 5 096 669, 1992.
19. G. Davis, I. R. Lauks and M. P. Zelin, Apparatus for assaying viscosity changes in fluid samples and method of conducting same, *United States Patent*, 5 628 961, 1997.
20. I. R. Lauks, Point-of-care in-vitro blood analysis system, *United States Patent*, 6 845 327, 2005.
21. V. Srinivasan, V. K. Pamula and R. B. Fair, An integrated digital microfluidic lab-on-a-chip for clinical diagnostics on human physiological fluids, *Lab Chip*, 2004, **4**, 310–315.
22. B. K. Russell, J. Woodward, W. Radtke and C. Dykes, System method and device for monitoring physiological parameters of a person, *United States Patent Application*, 2012/0165645, 2012.

23. M. Banet, T. Watlington and J. Moon, Body-worn vital sign monitor, *United States Patent Application*, 2011/0224500, 2011.
24. L. T. Antonelli and C.-L. Desjardins, Remote blood pressure waveform sensing method and apparatus, *United States Patent*, 8177 721, 2012.
25. D. A. Hull, Wireless pulse oximeter configured for web serving, remote patient monitoring and method of operation, *United States Patent Application*, 2005/0113655, 2005.
26. M. J. Banet, Vital signs monitor used for conditioning a patient's response, *United States Patent*, 7, 481 772, 2009.
27. Y. Geva, J. Clauser, V. Newberg, M. Jandes, M. Menard, G. Michelson, M. Petrucci and J. Van Schagen, Methods and apparatus for processing physiological data acquired from an ambulatory physiological monitoring unit, *United States Patent Application*, 2010/0249541, 2010.
28. E. Parton, Listen to your heart, *European Medical Device Technology*, 2012, 40–41.
29. J. M. Rueter, Method and apparatus for remote patient monitoring, *International Patent Application*, WO2007/060558, 2007.
30. Philips Remote Patient Monitoring: Easy for patients, efficient for clinicians, Philips Medical Systems, reference 4522 962 27751, 2007.
31. Intel Health Guide, PHS 6000, Intel Corporation, reference 319465 - 001 EN RSLT-A-0439, 2009.
32. G. Boyajian, D. Stern and B. Seitz, System and method for monitoring a site using time gap analysis, *United States Patent Application*, 2007/0195703, 2007.
33. S. J. Brown, Networked remote patient monitoring with handheld devices, *United States Patent*, 8 249 894, 2012.
34. C. Cronrath, F. Andreu and A. Val Vicente, Medical software download to mobile phone, *European Patent Application*, 1722310A, 2006.
35. H. Baldus and M. J. Elixmann, Mobile monitoring, *European Patent*, 1827214B, 2012.

CHAPTER 8

Block Copolymers

8.1 INTRODUCTION

The patent literature is vast, and over eight million United States patents have been granted since the first patent was awarded to Samuel Hopkins on 31 July 1790.[1] It is difficult to predict which patents will remain of general interest and which patents will produce new technologies and industries. Many patents are evolutionary, rather than revolutionary, and describe incremental improvements to existing processes or apparatus. The long time-scale, over a hundred years, for commercial exploitation of light-emitting diodes was described in Chapter 2. In contrast to this long time-scale, ground-breaking work by Cohen and Boyer.[2-4] on recombinant DNA technology at Stanford University in the 1970s underpins the worldwide bio-technology industry. Nowadays, Internet transactions, sensitive business and government communications are encrypted for protection using the so-called RSA codes based on the theory of prime numbers and congruence.[5] The RSA codes are named after the inventors and described in the patent literature.[6] Another area with potential for significant exploitation relates to block copolymers.

Exploring Materials through Patent Information
By David Segal
© David Segal, 2015
Published by the Royal Society of Chemistry, www.rsc.org

8.2 BLOCK COPOLYMERS

A block copolymer consists of two polymer chains or blocks which are covalently bonded to each other and which are chemically different.[7] An example is polystyrene-polymethylmethacrylate, which is derived from styrene and methylmethacrylate monomers and contains polystyrene and polymethylmethacrylate polymers. Two different polymer chains constitute a diblock copolymer, while a triblock copolymer has three different polymer chains. Ordered structures are produced in a block copolymer melt due to a balance between repulsion between dissimilar molecules favouring phase separation and entropic effects favouring mixing.[8] The blocks cannot separate completely because they are covalently bonded together and diblock copolymer chains are organised so that the individual blocks are on opposite sides of an interface.[7] When present in a thin film, oriented nanometre-scale patterns or domains are present in the copolymer.[7,9] Thus, lamellae align parallel or perpendicular to the substrate, cylinders align parallel or perpendicular to the substrate or spheres of one polymer phase are dispersed in the other polymer matrix. Cylinders and spheres have a colloidal dimension.

Phase separation in block copolymers can be described by analogy with a blend of two distinct polymers A and B.[7] The free energy for mixing A and B in a blend, ΔG_{mix}, is given approximately by the Flory–Huggins equation:

$$\frac{\Delta G_{mix}}{kT} = \frac{1}{N_A}\ln(f_A) + \frac{1}{N_B}\ln(f_B) + f_A f_B \chi \qquad (8.1)$$

where N_A and N_B represent the degree of polymerization, f_A and f_B represent the composition (mole or volume fractions) for A and B, while χ is the Flory–Huggins interaction parameter, T is the absolute temperature and k is the Boltmann constant. The first two terms in Equation 8.1 relate to configurational entropy for the system and are controlled by the polymerization chemistry to change the relative lengths of the chains and fractions of A *versus* B. The third term, $f_A f_B \chi$ is affected by the chemistry of the molecules and temperature.

Formation of ordered structures in block copolymers is often referred to in the semiconductor industry as directed self-assembly. These materials have attracted attention in this industry as they offer a method for producing patterns on integrated circuits with smaller dimensions and higher densities than are currently available (Figure 8.1).[10,11] The patterns represent interconnect lines that are constantly being decreased as integrated circuits are reduced in size. This reduction in size is due, in part, to demand for increased portability, computing power, memory capacity and energy efficiency in devices such as smartphones and tablet computers as well as other consumer goods.[12]

8.3   SEMICONDUCTOR DEVICES

Semiconductor devices have been described by Cheng *et al.*[10] These devices contain a network of circuits on a substrate, often with several layers of circuit wiring with interconnects used to connect the layers to each other and to underlying transistors. Vias or contact holes are formed and connect with other layers, after which they are filled in with a metal to form interconnects so that the various layers of circuitry are in electrical communication with each other. Methods for forming inter-connects

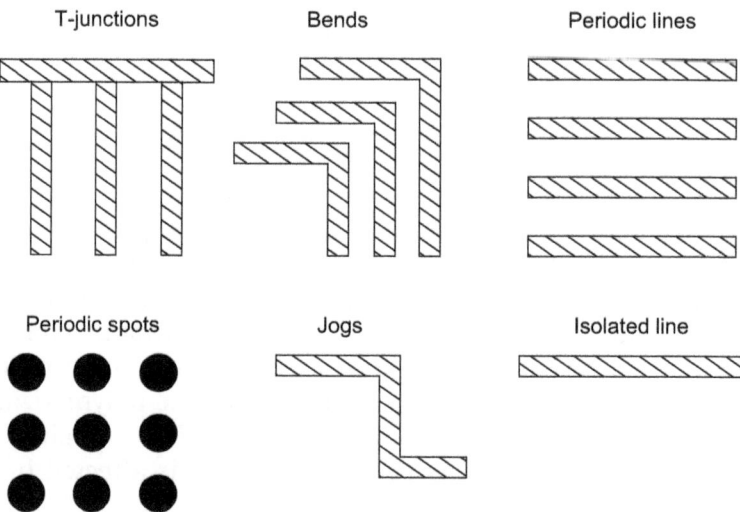

Figure 8.1   Schematic diagram showing examples of interconnect lines derived from lithographic patterns.[11]

require a series of lithographic and etching steps to define the positions and dimensions of the vias that define the positions and dimensions of the corresponding interconnects. Features formed from conventional optical lithography for volume manufacture have reached the resolution limit for existing lithographic tools. These features are produced by using an argon fluoride excimer laser with a 193 nm wavelength. While both electron beam and extreme ultra-violet lithography have the potential for higher resolution, they are not sufficiently developed for high-volume manufacture. Directed self-assembly of block copolymers has the potential for increasing the resolution of existing lithographic tools for semiconductor fabrication.

8.4 DIRECTED SELF-ASSEMBLY AND LITHOGRAPHY

Figure 8.2(a) shows a substrate with a trench.[13] In Figure 8.2(b) a self-assembled A-B block copolymer with lyophilic A blocks and lyophobic B blocks is deposited into the trench to form a layer with alternating stripes of A and B domains. In Figure 8.2(c), type A domains have been removed by selective chemical etching leaving type B domains to form a relief pattern in the trench that can act as a template for patterning the bottom surface. Block copolymer films are annealed or cured, typically for 24 hours at around 180 °C. Endou and Sasago showed that if annealing was carried out when the copolymer film was covered with a water-soluble polymer film such as polyvinyl alcohol then annealing times were reduced to six hours or less.[14] They suggested that hydrophilic groups in the polymer film aided the annealing process through interaction with domains in the self-assembled polymer. Millward produced sub-lithographic features, that is, conductive lines with widths of 57.5 nm or less by standard lithographic methods combined with directed self-assembly of block copolymers while Russell *et al.*[15,16] prepared a polymeric replica of a patterned crystalline surface onto which a block copolymer was spin-coated and annealed. As the polymer replica is flexible, roll-to-roll processing methods are feasible for fabrication of high-density arrays. Silicon single crystals with lateral spacings of less than 100 nm can be used for producing

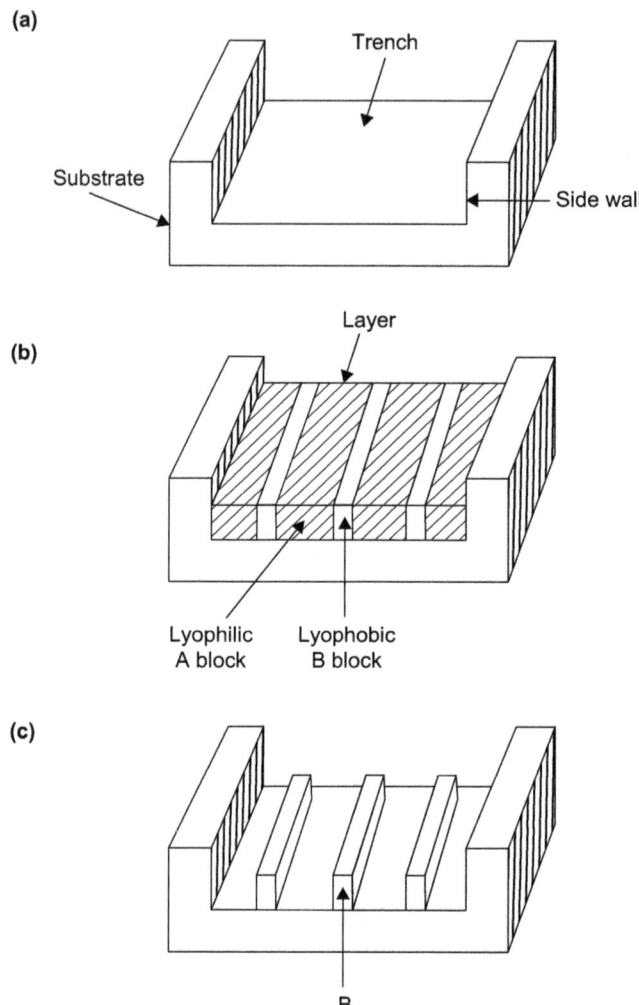

Figure 8.2 Self-assembly of A-B block copolymer.[13]

the polymer replica. The process is shown schematically in Figure 8.3.

8.5 SUMMARY

Phase separation of block copolymers, also known as directed self-assembly has potential application for increasing the

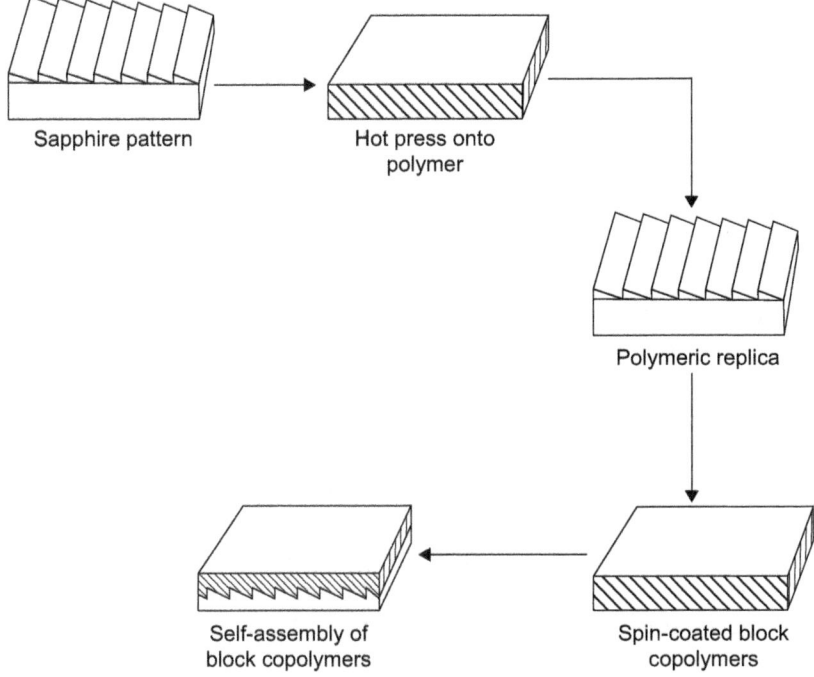

Figure 8.3 Formation of polymeric replica from a patterned substrate.[16]

resolution of existing lithographic tools to increase the density of patterns and interconnects in integrated circuits.

REFERENCES

1. S. Hopkins, Improvement in the making of Pot Ash and Pearl Ash by a new apparatus and process, *United States Patent*, X000001, 1790.
2. S. N. Cohen and H. W. Boyer, Process for producing biologically functional molecular chimeras, *United States Patent*, 4 237 224, 1980.
3. S. N. Cohen and H. W. Boyer, Biologically functional molecular chimeras, *United States Patent*, 4 468 464, 1984.
4. S. N. Cohen and H. W. Boyer, Biologically functional molecular chimeras, *United States Patent*, 4 740 470, 1988.

5. M. Liebeck, *A concise introduction to pure mathematics*, CRC Press, 2011, pp. 123–126.
6. R. L. Rivest, A. Shamir and L. K. Adleman, Cryptographic communications system and method, *United States Patent*, 4 405 829, 1983.
7. R. A. Segalman, Patterning with block copolymer thin films, *Materials Science and Engineering*, 2005, **R48**, 191–226.
8. E. R. Dufresne, H. Noh, V. Saranathan, S. G. J. Mochrie, H. Cao and R. O. Prum, Self-assembly of amorphous biophotonic nanostructures by phase separation, *Soft Matter*, 2009, 5, 1792–1795.
9. A. Knoll, A. Horvat, K. S. Lyakhova, G. Krausch, G. J. A. Sevink, A. V. Zvelindovsky and R. Magerle, Phase behaviour in thin films of cylinder-forming block copolymers, *Phys. Rev. Lett.*, 2002, **89**(3), 035501-1–035501-4.
10. J. Cheng, Y.-H. Na, K. Lai, C. T. Rettner, D. P. Sanders and W.-K. Li, Directed self-assembly of block copolymers using segmented prepatterns, *European Patent*, 2379441B, 2012.
11. Y.-C. Tseng and S. B. Darling, Block copolymer nanostructure for technology, *Polymers*, 2010, 2, 470–489.
12. G. Sandhu, Pitch multiplication using self-assembling materials, *United States Patent*, 7 923 373, 2011.
13. R. Koole, J. Dijksman, S. Wuister and E. Peeters, Lithography using self-assembled polymers, *International Patent Application*, WO 2012/031818, 2012.
14. M. Endou and M. Sasago, Method of accelerating self-assembly of block copolymer and method of forming self-assembled pattern of block copolymer using the accelerating method, *United States Patent Application*, 2011/0186544, 2011.
15. D. B. Millward, Methods using block copolymer self-assembly for sub-lithographic patterning, *United States Patent*, 7 964 107, 2011.
16. T. P. Russell, S. Park, D. H. Lee and T. Xu, Self-assembly of block copolymers on topographically patterned polymeric substrates, *United States Patent*, 8 247 033, 2012.

CHAPTER 9
Aerogels

9.1 INTRODUCTION

Porosity is an important physical property for advanced ceramics as it can have either advantageous or deleterious effects on performance depending on their application. Whereas traditional ceramics are derived from naturally occurring raw materials such as clay minerals, advanced ceramics are developed from chemical synthetic routes or from naturally occurring materials that have been highly refined. Sintered structural ceramic components require theoretical or near-theoretical density because pores act as sources of flaws and lower their fracture strength. Finely divided metals in three-way vehicle exhaust catalysts are dispersed on high surface area porous coatings, for example alumina, for the conversion of carbon monoxide, hydrocarbons and nitrogen oxides to carbon dioxide, water and nitrogen. Molecular separations in high-pressure liquid chromatography are carried out on packed columns of porous oxide microspheres. Synthetic membranes, both inorganic and polymer are used to purify materials in diverse areas including food and liquid effluent processing and water desalination by reverse osmosis. Pore size is important in gas separation as it affects the type of transport mechanism across the membrane, namely Knudsen diffusion, laminar flow or surface diffusion.

Exploring Materials through Patent Information
By David Segal
© David Segal, 2015
Published by the Royal Society of Chemistry, www.rsc.org

Table 9.1 General properties of aerogels.[1]

General properties of aerogels
• Approximately 95% open interconnected porosity
• Nanometre-sized pores
• High surface area
• Low density, extremely lightweight
• Low thermal conductivity for thermal insulation
• Optical transparency (for silica aerogels) due to minimal light scattering from pores in the visible spectrum
• Thermal stability
• Low sound velocity useful for soundproof materials

Porous aluminosilicate zeolites have been used to replace polyphosphates in detergents as an adsorbent for calcium ions from wash liquors. Calcium hydroxyapatite, $Ca_{10}(PO_4)_6(OH)_2$ has a structure similar to that of naturally occurring bone and tooth minerals and is biocompatible with them. It has applications as porous coatings on the stems of metallic hip implants. There is also increasing interest nowadays in porous insulating materials for both commercial and residential uses in order to reduce both energy bills and carbon emissions. Low thermal conductivity is a desirable property of insulating materials. Aerogels are another class of porous materials with nanometre-sized pores and they can be described as solid foams. Their general properties are summarised in Table 9.1.[1]

9.2 SUPERCRITICAL CONDITIONS

A pressure difference Δp exists across the liquid-air interface for menisci in the pores of a drying gel and is given by the Laplace equation,[2]

$$\Delta p = 2\gamma \cos \theta / r_p \qquad (9.1)$$

where r_p is the pore radius, γ is the surface tension of liquid and θ is the contact angle at the solid-liquid-air boundary. Figure 9.1 shows the variation of this capillary pressure with pore radius for water assuming perfect wetting ($\cos \theta = 1$).[3] The effect of the capillary pressure is a stress acting on the walls of pores in the gel whose magnitude increases with decreasing pore size and which can lead to the collapse of capillaries and crack formation in the dry gel. The term gel can refer to both aquagels

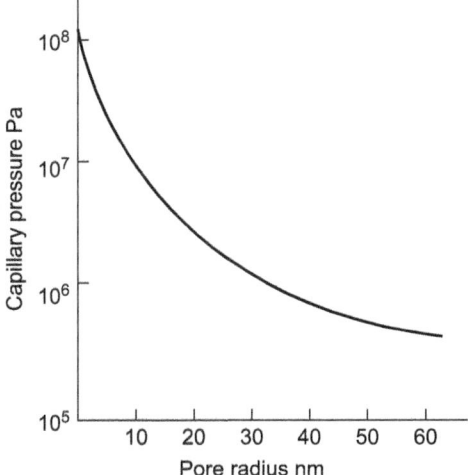

Figure 9.1   Variation of capillary pressure with pore radius for a water-filled capillary.[3]

and alcogels.[4] Aquagels, also known as hydrogels, are solids derived from aqueous sodium silicate solutions in a process that can be represented by polymerisation of silicic acid, $Si(OH)_4$. However, alcogels are monolithic materials derived from hydrolysis of an alkoxide such as tetraethoxysilane in a sol-gel process and have the same volume as the alkoxide reagent. The alkoxide is known as the sol and polymerisation of hydrolysis products results in gel formation in the form of a rigid monolith, fibres, coatings or powders. This sol-gel transition is irreversible. However, the phrase sol-gel can be confusing because a sol is also a colloidal system in which particles with at least one dimension such as a diameter in the range 1 nm and 1 μm is dispersed in a liquid medium. In sol-gel processing of colloids an aqueous sol is dried to a stiff gel in the form of spheres, fibres, fragments or coatings and this sol-gel transition is usually reversible.

Note that oxides made by heating hydrogels are referred to as xerogels when completely dehydrated.

Stresses responsible for cracking of gels disappear if liquid menisci in pores are eliminated and this can be achieved by using supercritical drying conditions in an autoclave.

A supercritical fluid is a substance above its critical temperature where it remains as a single fluid phase. Increase in pressure affects fluid density but does not produce a separate fluid phase. Examples of critical temperatures and pressures are shown in Table 9.2 and solids made by supercritical drying of alcogels and aquagels are known as aerogels.[5] Carbon dioxide can also act as a supercritical fluid with a critical temperature and pressure of 304.1 K and 7.38 M Pa, respectively,[6] which is considerably lower than for water (Table 9.2). The phase diagram for carbon dioxide is shown in Figure 9.2 to indicate the supercritical region.[6]

Table 9.2 Critical parameters for water and non-aqueous solvents.[5]

Solvent	Boiling point K	Critical pressure M Pa	Critical temperature K
Acetone	329	4.7	508
Methanol	336	8.0	513
Ethanol	351	6.4	513
1-Propanol	370	5.2	538
1-Butanol	390	4.4	563
Ethylene glycol	470	7.7	643
Water	373	21.9	648

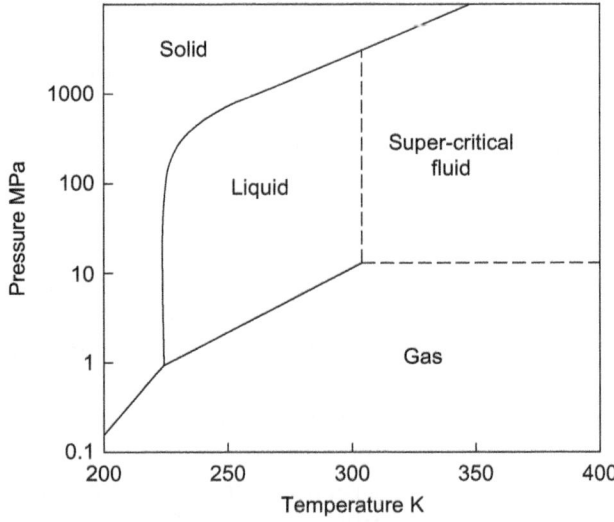

Figure 9.2 Phase diagram for carbon dioxide.[6]

The method for producing aerogels was devised by Kistler,[7] in which water was displaced from a hydrogel by alcohol before placing the alcogel in an autoclave and removing vapour under supercritical conditions. It is necessary to substitute alcohol for water because inorganic oxides are soluble in aqueous media at the temperatures and pressures used for drying, but preparation of alcogels directly from alkoxides avoids this solvent exchange process.

9.3 AEROGELS AND THEIR APPLICATIONS

9.3.1 Organic Aerogels

Aerogels derived from silica, alumina or other metal oxides are described as inorganic. Organic aerogels are derived from organic compounds with organic linking groups. Polyisocyanate-based rigid foams, for example, polyurethane and polyisocyanate foams are used as thermal insulation in refrigerators.[8] The foams are prepared by reacting a polyisocyanate and polyol in the presence of a blowing agent. The thermal insulating properties of rigid foams depend on a number of factors including the cell size of the foam and the thermal conductivity of the blowing agent. Conventional blowing agents have been fully halogenated chlorofluorocarbons, in particular trichlorofluoromethane, as the latter has a low thermal conductivity. Concern over the role of chlorofluorocarbons on the depletion of ozone in the atmosphere has led to a search for alternative blowing agents. In an alternative approach a polymeric diphenylmethane diisocyanate was first blended with dichloromethane after which a trimerisation catalyst was added.[8] A rigid gel formed after about 8 hours and supercritically dried at 323 K and 9.5 M Pa. The resulting opaque aerogel had a density of 0.35 g cm^{-3}, a surface area of 210 m^2 g^{-1} and an average pore diameter of 18 nm. As aerogels are filled with air rather than a blowing agent they are environmentally friendly.

During the drying process a gel derived from polyisocyanate shrinks leading to an increase in density. When carbon black was added to a mixture of polyisocyanate, dichloromethane and trimerisation catalyst before gelation and treatment with supercritical carbon dioxide, the resulting aerogel had a density of

0.187 g cm^{-3} compared to 0.355 g cm^{-3} for an aerogel that did not contain carbon black and a pore size of 25.8 nm compared to 8.9 nm when carbon black was absent.[9]

Organic aerogels derived from base-catalysed polymerisation reactions between resorcinol (1,3 dihydroxy benzene) and formaldehyde and also the reaction between melamine and formaldehyde have been summarised.[10] A modification of the reaction between resorcinol and formaldehyde has been carried out, in which thio-containing molecules, amine-containing-molecules and nitro-containing molecules have been used instead of resorcinol and the reaction schemes are shown schematically in Figure 9.3.[10] Aerogels are dried either by repeated solvent exchange or by use of supercritical drying and these functionalised derivatives of resorcinol appeared to aid the drying process.

A further example of an organic aerogel involves a polymer formed between an aryl alcohol, an aldehyde and a polyol.[11] For example, the reaction between resorcinol, formaldehyde and pentaerythritol produced a solid gel after which solvent, acetonitrile, was exchanged with acetone and the latter was removed by use of supercritical carbon dioxide. The aerogel had a density of 0.30 g cm^{-3}, surface area 493 m^2 g^{-1}, pore size 18.2 nm and thermal conductivity 0.023 W m^{-1} K^{-1}. A maleinised polybutadiene was produced by adding maleic anhydride to a polybutadiene that was then converted to a gel on addition of a triethylamine catalyst and a hardener in an acetone solution.[12] Acetone was displaced by carbon dioxide and an aerogel produced at supercritical conditions in carbon dioxide. The aerogel had a surface area of 5 m^2 g^{-1} and thermal conductivity of 0.023 W m^{-1} K^{-1} at 303 K. A composite material was produced by adding quartz fibre to maleinised polybutadiene before gelation. The composite aerogel had a density of 0.1695 g cm^{-3} and thermal conductivity of 0.025 W m^{-1} K^{-1}.

9.3.2 Carbon Aerogels

Carbon aerogels are produced when organic aerogels are pyrolysed at high temperature. The use of a carbon aerogel electrode loaded with a noble metal catalyst such as platinum has been suggested as electrodes for fuel cells.[13] When the electrode had a

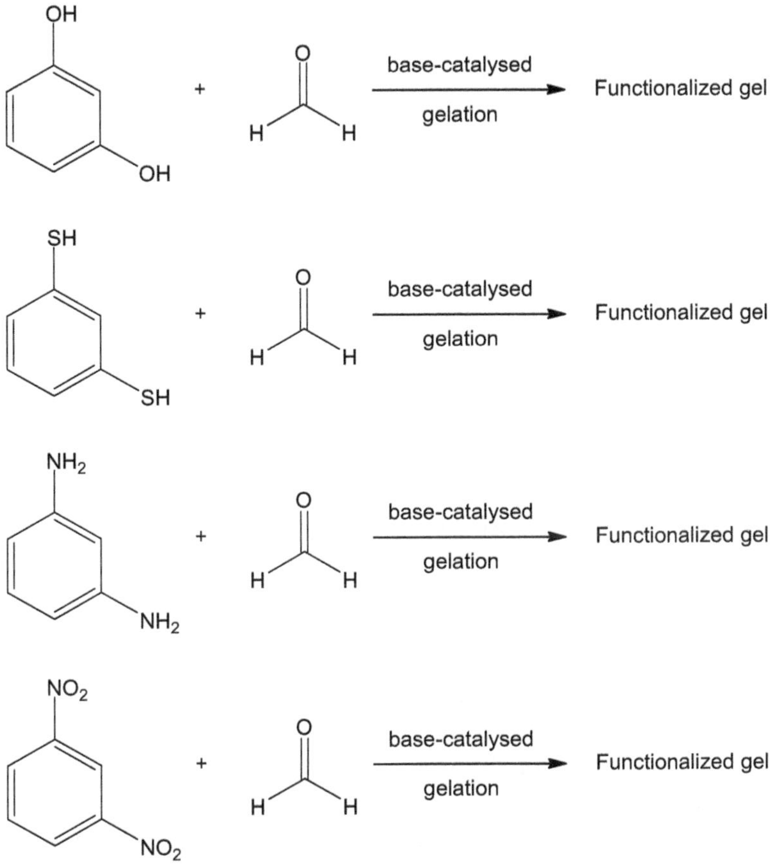

Figure 9.3   1,3-difunctionalised benzene derivatives and gels derived from them.[10]

surface area in the range 400–1200 m^2 g^{-1}, a density between 0.3 and 1.2 g cm^{-3} and a reactant/catalyst ratio of about 50–400, it overcame high polarisation resistance that limited power capability of the cell. An aerogel prepared from the condensation of resorcinol and formaldehyde in the presence of sodium carbonate and dried by supercritical carbon dioxide had a surface area of 888 m^2 g^{-1}, pore volume 1.24 cm^3 g^{-1} and average pore diameter 5.5 nm.[14] Pyrolysis of the aerogel at 1273 K produced a carbon aerogel with a surface area of 741 m^2 g^{-1}, pore volume of 0.77 cm^3 g^{-1} and average pore diameter 3.4 nm. Aerogel pieces were placed in an autoclave with a fine powder of dimethyl (cyclooctadiene) platinum and exposed to supercritical carbon dioxide. After releasing the gas pressure the aerogel pieces were pyrolysed to produce a carbon aerogel with a platinum content of 7 weight %. The metal-doped carbon aerogel was considered useful for fuel cell electrodes.

Carbon aerogel pieces with an average pore size of 25 nm and a pore volume of 1.3 cm^3 g^{-1} were soaked in a nickel nitrate solution in acetone, removed and dried at 373 K in an oven and then heated in a reducing atmosphere at 773 K to form an aerogel containing 2.3 weight % nickel.[15] The Ni-doped carbon aerogel was then heated in a sealed tube at 1123 K in the presence of magnesium. Chemical analysis of the carbon aerogel showed that it contained 16.4 weight % of magnesium that was not oxidised. In this approach, nickel does not wet the surface of the pores in the carbon aerogel but wets the surface of magnesium. Potential applications for the metal-filled aerogels were battery electrodes and catalyst supports. Another application for carbon aerogels is in carbon composite materials.[16] For example, a solution of resorcinol, formaldehyde and sodium carbonate solution in water was mixed with carbon fibres and after gelation the composite gel was immersed in acetone to displace water from the gel that was then subjected to supercritical drying by carbon dioxide at 313 K and 6 M Pa for 6 hours. The dried composite was then pyrolysed at 1323 K to form a glassy-carbon composite monolith.

Cryogenic pumps have many applications, for example, in particle accelerators, ion implantation and thin film deposition and operate at very low temperatures around 70 K or even 15 K.[17] Sorbents are used to adsorb gas from a pumping system but in order to regenerate the sorbent it must be removed from the pumping system and heated or 'baked' until adsorbed molecules are removed. If the sorbent has a low thermal conductivity, as carbon aerogel has, then the time required to cool the sorbent during pumping and to bake the sorbent will be longer than a material with a higher thermal conductivity. Composite aerogel/ metal sheets in which carbon aerogel is bonded onto panels of metal mesh or foil approximately 0.12 cm thick have found application in pumping systems as the increase in thermal conductivity due to the metal panels reduces the time to cool down the sorbent during pumping and to heat the sorbent during regeneration.

9.3.3 Monolithic Aerogels

Preparation of transparent monolithic aerogels is difficult as aerogels are prone to fracture and cracking on drying. The following method has been developed for metal oxide aerogels.[18] For example, a mixed solution of tantalum ethoxide and tetraethoxysilane was prepared in ethanol and then added to an aqueous solution of tetrafluoroboric acid. The mixed hydrolysed alkoxide solutions are allowed to undergo condensation reactions and gel in a sealed container that does not have a gas-tight seal. The containment vessel is kept in an autoclave and water is placed around the containment vessel in the autoclave. The autoclave is heated to 573 K and the water atmosphere is kept at 13.5 M Pa so that ethanol passes from the containment vessel into the autoclave. The resulting aerogel had a density of 0.35 g cm^{-3} and a surface area of 700 m^2 g^{-1}. Transparent transition metal oxide aerogels could be prepared with densities from 0.005 g cm^{-3} to 0.4 g cm^{-3} with surface areas up to about 750 m^2 g^{-1} and potential optical, magnetic and catalytic applications.

9.3.4 Hydrophobic Aerogels

Lightweight silica aerogels with low thermal conductivity are an attractive choice for thermal insulation, provided they are hydrophobic as the aerogel microstructure collapses on adsorption of water from the liquid or vapour phase. This requirement for hydrophobicity has attracted attention in the patent literature. Thus hydrogel beads were produced by addition of sulphuric acid to sodium silicate solution (water glass solution).[19] The hydrogel was washed with deionised water to remove entrained salts and residual water in the hydrogel was then displaced by isopropanol. The resulting alcogel was heated in the presence of isopropanol in an autoclave to 523 K at 9 M Pa over a period of 5 hours. The resulting silica aerogel was ground to a powder that had a density of 0.11 g cm^{-3}, a surface area of 380 m^2 g^{-1} and thermal conductivity of 0.028 W m^{-1} K^{-1}. The hydrophobic nature of the aerogel is illustrated by its water uptake of 1.4 weight % compared to 38.6 weight % for an aerogel that did not have the extensive rinsing procedure. Potential applications for the hydrophobic aerogel include additives for rubber and catalyst supports as well as thermal insulation. Another approach was to mix a water glass solution and acidic ion exchange resin to lower the pH of the solution to 2.4.[20] A gel was obtained on raising the pH to 4.7 with sodium hydroxide. The hydrogel was washed with a diether such as dimethoxyethane to displace entrained water, after which the gel was silylated with, for example, trimethylchlorosilane at room temperature. The gel was then washed again with dimethoxyethane and dried over a period of 17 hours. The resulting transparent hydrophobic aerogel had a density of 0.13 g cm^{-3}, surface area of 500 m^2 g^{-1} and thermal conductivity of 0.015 W m^{-1} K^{-1}. A feature of the synthesis is that the ether did not react with the silylating agent.

An alternative method for producing a hydrophobic silica aerogel involved first forming an alcogel from an alkoxysilane solution such as tetramethoxysilane in ethanol using aqueous ammonia as a catalyst.[21] The gel was dried under supercritical conditions of 353 K and 16 M Pa in an autoclave containing ethanol and carbon dioxide and a hydrophobising agent, an organosilane such as hexamethyldisilazane. A transparent hydrophobic aerogel 5 cm diameter and 0.5 cm thick was produced with a surface area of 1108 m^2 g^{-1} and density of 0.04 g cm^{-3}.

Hydrophobic aerogels derived from fluorine-containing compounds have the potential to remove organic liquids from aqueous solutions.[22] As an example, an alcogel was produced from the hydrolysis and condensation reactions on addition of aqueous ammonia to a solution of trifluoropropyl-trimethoxysilane in methyl alcohol. Alcogels were dried either under supercritical conditions to extract the alcohol solvent or slowly dried by evaporation over a period up to 28 days. Aerogels were in granular or powdered form and the degree of hydrophobic nature of the alcogel was measured using a sessile water droplet method by measuring the angle of the droplet relative to a flat surface. Alcogels gave a contact angle greater than 90°, characteristic of a hydrophobic material. Table 9.3 shows a comparison of the adsorption capacities for four solvents using hydrophobic silica aerogel and granulated carbon granules illustrating the enhanced capacity for the aerogel.

Table 9.3 Comparison of adsorption capacities of hydrophobic silica aerogel and granulated activated carbon.[22]

Solvent	Hydrophobic silica aerogel (gram per gram)	Granulated activated carbon (gram per gram)
Toluene	0.833	0.026
Cyclohexane	0.458	0.011
Trichloroethylene	11.89	0.091
Ethanol	1.94	0.028

Another method for producing a hydrophobic silica aerogel containing surface fluorine atoms is to first dehydrate the aerogel by heating at 573 K followed by its exposure to a vapour of trifluoropropyl-trimethoxysilane.[23] Although aerogels are attractive candidates for thermal insulation applications, they are brittle solids and difficult to handle. Mechanically robust aerogel monoliths were obtained by coating their surfaces with isocyanates and allowing the coating to cure at room temperature.[24] The reaction of a silica aerogel with an isocyanate to form a coating on the aerogel monolith is shown schematically in Figure 9.4.

A hydrophobic solution for coating an aerogel was prepared by mixing a solution of poly (methyl methacrylate) with a dispersion of silica particles derived from an aerogel that had surface functional groups derivatized with a silylating agent such as trimethylchlorosilane.[25] Contact angles on coated monoliths were greater than 150°, for example, 159° and angles greater than 150° were designated as characteristic of superhydrophobic coatings. A potential application for hydrophobic aerogels is in the encapsulation and slow release of drugs.[26] Alcogels with a diameter of 1.4 cm and a height of 0.5 mm were obtained by hydrolysis and condensation of tetraethoxysilane, aged in ethyl alcohol for 3 days to increase the mechanical strength of the alcogel, immersed in a solution of the photoinitiator eosin-Y and dried by supercritical carbon dioxide at 313 K and 10.3 M Pa. Hydrophobic aerogels were prepared by reacting the eosin-Y functionalised aerogel with hexamethyldisilazane with supercritical carbon dioxide as the solvent at 20.7 M Pa and 333 K. The eosin-loaded hydrophobic aerogel was immersed in a mixed solution of polyethylene glycol and a diacrylate polymer and a hydrogel coating of polyethylene glycol was obtained on

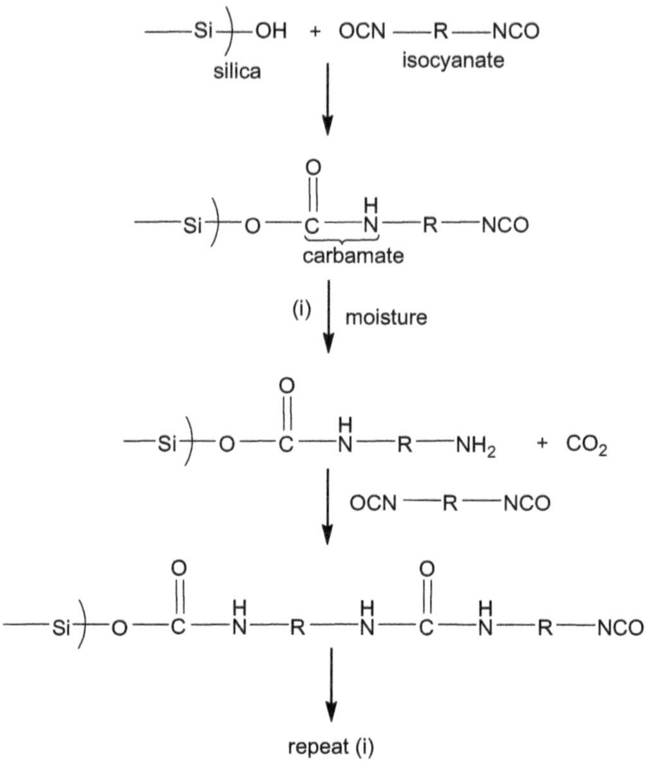

Figure 9.4 Schematic diagram showing the interaction of the surface of a silica aerogel and in isocyanate coating medium.[24]

photopolymerisation in visible light. Thus, the outer hydrogel layer of the composite material could encapsulate a hydrophilic pharmaceutical compound while a hydrophobic drug is incorporated into the inner hydrophobic core, both drugs having different rates of release.

An application of silica aerogels for thermal insulation that does not require hydrophobic material relates to light-emitting diodes (LED) for lighting systems.[27] The LED is usually encapsulated in a transparent resin for protection against dust and water vapour and phosphor powders can also be embedded throughout the resin. Phosphors absorb light emitted from the LED and emit radiation of a lower wavelength with the desired optical properties. However, heat generated by the LED can affect emission from the phosphors but placing a transparent aerogel layer up to 20 μm between the LED and resin shields the phosphors from the high temperature generated by the LED chip.

9.3.5 Sub-critical Drying of Aerogels

Sub-critical drying of aerogels takes place below both the critical temperature and critical pressure of the drying medium, for example, by use of microwaves in a standard microwave oven.[28] As an illustration of the technique a solid alcogel formed by hydrolysis of a tetraethoxysilane solution in ethyl alcohol was placed in n-hexane to displace the alcohol from the alcogel. A hydrophobic alcogel was produced by addition of trimethyl-chlorosilane to the gel, after which residual silane was displaced from the pores by n-hexane. Gels that contain an organic solvent have been referred to as lyogels.[28] Gel fragments were placed in a microwave oven under a nitrogen atmosphere and subjected to microwave radiation at a frequency of 2.450 ± 25 M Hz for 40 minutes. The resulting aerogel had a density of 0.14 g cm^{-3}, while a gel that was dried for the same time in air had a density of 0.76 g cm^{-3}. Another approach involves treating an alcogel with a liquid that has a temperature above the boiling point of the pore liquid.[29] For example, a silica sol was first produced by passing a sodium silicate solution through an ion exchange column to lower the pH to between 2.5 and 2.9 and then raising the pH of the sol to about 5 by controlled addition of sodium hydroxide that caused gelation. Cylindrical hydrogels were first flushed repeatedly with acetone that was then displaced from the gel by rinsing with n-heptane. A hydrophobic gel was produced on treatment with trimethyl chlorosilane. The wet gel was then immersed in boiling water, after which the entrained heptane evaporated and in the process the gel disintegrated to a fine powder that rose to the water surface. The dry fluffy gel powder had a surface area of 600 m^2 g^{-1} and density of 0.08 g cm^{-3}. A further sub-critical drying method involved rapidly drying gel granules containing pore liquid by raising the temperature of a solvent in which the gel was immersed to above the boiling point of the liquid at atmospheric pressure but by keeping the pressure below the supercritical pressure of the solvent.[30] Pore liquid was typically removed in about 20 seconds.

9.3.6 Aerogel Coatings and Fibres

Anti-reflection coatings have been produced by physical vapour deposition such as sputtering and by wet chemical techniques, for example, sol-gel processing.[4,31] The anti-reflection coating should have a refractive index smaller than the substrate and for glass lenses the refractive index is approximately 1.5. Magnesium fluoride with a refractive index of 1.28 has been used as an anti-reflection coating and deposited by physical vapour deposition but a lower index is required to produce a reflectance of less that 1% in the visible region. The transparency, low density and re-fractive index of around 1.1 make silica aerogels attractive candidates for anti-reflection coatings.[32] Two sols were prepared by hydrolysis of tetraethoxysilane, the first by base hydrolysis with aqueous ammonia after which the sol was acidified by addition of hydrochloric acid.[31] Silica particles in this acidic sol had a size in the range 10–50 nm. Another acidic sol was prepared by addition of hydrochloric acid directly to tetraethoxysilane in ethyl alcohol and the particle size in this sol was 2 nm or less. The two sols were combined to give a bimodal particle size distribution and a silicon wafer was dip-coated with the mixed sol, dried at 353 K and heated further at 433 K to produce a silica aerogel film. The latter had a thickness of 146 nm and refractive index of 1.24. A modified technique produced a tough water-repellent anti-reflection coating with a refractive index between 1.2 and 1.3.[32] Here an alcogel derived from tetraethoxysilane was dispersed in methyl isobutyl ketone and a solution of trimethylchlorosilane added to the dispersion to produce a hydrophobic sol-like silica. The latter was mixed with an ultraviolet-curable resin solution, dip-coated onto a silicon substrate and then dried and baked at 423 K for 1 hour to form a silica aerogel coating.

Aerogels are brittle materials and their exploitation is dependent in part on their ease of handling. In order to fabricate a fibrous silica aerogel sheet, a hydrophilic gel was first prepared from a sodium silicate solution by addition of ethyl alcohol.[33] The surface of the gel was made hydrophobic by treatment with trimethylchlorosilane. A sample of the hydrophobic gel and polymethyl methacrylate were dissolved in *N,N*-dimethylformamide and this solution was used for electrospinning a fibre web in the form of a sheet by ejection from a spinning nozzle under an applied electric field. The sheet contained silica aerogel in fibrous form and polymethyl methacrylate was removed from the sheet on heating the latter at 573 K but the fibrous structure collapsed to yield grains and lumps in the sheet. Fibrous aerogel sheets had potential applications as adsorbents and thermal insulation. Silica aerogels have also been used as cladding on optical fibres.[34] A quartz fibre was dip-coated into a hydrolysed tetramethoxysilane solution and subjected to a hydrophobic treatment with a hexamethyldisilazane solution in ethyl alcohol. The alcohol in the pores was displaced by carbon dioxide and dried under supercritical conditions at 313 K and 8 M Pa. The silica aerogel had a density of 0.10 g cm^{-3}, particle size of 2–3 nm and refractive index of 1.03.

9.3.7 Carbon Nanotube and Graphene Aerogels

Aerogels are often associated with silica-based systems, perhaps because of the early work carried out by Kistler and many examples of silica aerogels have been described in this chapter.[7] However, aerogels are networks of nanoparticles and there is no inherent reason why aerogels cannot be produced from nanoparticles of other materials. Aerogels have been produced from carbon nanotubes with a preferred diameter in the range 10–15 nm.[35] For example, carbon nanotubes were first oxidised by a mixture of concentrated sulphuric and nitric acid and then washed with sodium hydroxide, then with water and finally repeatedly washed with dimethylformamide and dispersed in dimethylformamide. 1, 3-dicyclohexylcarbodiimide was added to catalyse the esterification reaction between alcohols and acid groups of the oxidised nanotubes. 2,2,2-trifluoroethanol was added to the gel to make the nanotube surfaces hydrophobic and increase the contact angle between nanotubes and solvent to prevent the collapse of the porous gel structure on drying. Solvent exchange between dimethylformamide in the pores with acetone and then between acetone and n-heptane resulted in heptane-filled pores. This solvent exchange was carried out in two steps as dimethylformamide and n-hexane are immiscible and acetone is miscible with both solvents. Small cylindrical aerogels, 0.2 cm in diameter, were obtained on evaporation of n-heptane at room temperature and had a density of 0.3 g cm^{-3}.

In a different preparation of carbon nanotube aerogels, a potassium salt of naphthalene made by addition of potassium metal to naphthalene in tetrahydrofuran (THF) was added to a sample of carbon nanotubes, a process that reduced the nanotubes producing a polyelectrolyte salt of nanotubes.[36] The salt was dispersed in dimethyl sulphoxide and this solution was frozen, after which an aerogel was recovered by freeze-drying. In this synthesis the formula of the polyelectrolyte salt was considered to be $K(THF)_jC_k$ where j is between 0 and 8 and k is an integer between 6 and 200. Aerogels could be prepared with surface areas up to 2000 m^2 g^{-1} and densities as low as 0.0001 g cm^{-3}. A graphene aerogel was prepared by first treating a dispersion of graphite oxide containing NH_3BF_3 in solution to a hydrothermal treatment at 453 K for 12 hours.[37] The resulting graphene aerogel was doped with 0.1 to 6 weight % of nitrogen and 0.1 to 2 weight % of boron and was recovered by freeze-drying. Graphene aerogels had surface areas ranging from 200 to 1000 m^2g^{-1}, densities between 0.02 and 0.05 g cm^{-3} and electrical conductivities between 0.0001 and 1 S cm^{-1}. The capacitance of aerogels containing both dopants was higher than for aerogels containing either nitrogen or boron and a possible application for the doped aerogel was for supercapacitors. The variety of materials that can be produced as aerogels can be illustrated by the formation of cellulose aerogels with densities as low as 0.001 g cm^{-3} and containing at least 50 weight % of cellulose nanocrystals.[6]

9.3.8 Aerogel Composites

Practical applications of aerogels are limited by their brittle and fragile nature but composite materials containing aerogels can overcome these limitations. Hollow glass spheres[38] were incorporated into an aerogel formed by the base-catalysed condensation between resorcinol and formaldehyde without the requirement for a binder. The effect of the hollow spheres and absence of binder was to produce a composite material with low thermal conductivity up to 0.5 W m^{-1} K^{-1}. The composites are useful as mould materials for metal casting because it is difficult for the casting material to spread in thin-walled zones of the mould for higher values of thermal conductivity. Silica aerogel

beads were blended with a thermoplastic, for example polyamides (nylon), polycarbonates or polyesters by melt extrusion.[39]
The addition of aerogel beads at concentrations as low as
5 weight % resulted in a composite that was more elastic and less
brittle at cryogenic temperature than the unmodified polymer.
Aerogel beads lower the thermal conductivity of the composite,
which is particularly useful for wood/thermoplastic polymer/
aerogel composites for outdoor decking planks, which can be hot
to the touch in sunlight with an increase in their thermal conductivity. Composites for thermal insulation were obtained by
soaking a fibreglass mat in a hydrolysed alcoholic solution of
tetraethoxysilane and supercritically dried in an atmosphere
of ethyl alcohol.[40] Fibres used for textiles are often composed of
many twisted single-fibre strands that have voids between the
single strands.[41] These voids can be filled with powdered aerogel
nanoparticles with diameters between 1 and 500 nm and the fine
powders adhere firmly to the strands. Fabrics made from the
fibres can be waterproof, fire resistant and thermally insulating.
Aerogel-foam composites with low thermal conductivity,
approximately 0.02 $W\ m^{-1}\ K^{-1}$ and low compressive strength
around 19 M Pa have been prepared.[42] For example, a polymer
foam such as polyurethane or polycarbonate with an open cell
structure is impregnated with precursor solutions suitable to
produce an aerogel was dried under supercritical conditions for
carbon dioxide.

9.3.9 Aerogels as Catalyst Supports

Desirable properties of catalyst materials include high surface
area, selectivity, thermal stability against loss of surface area and
ease of handling.[43] Catalyst materials are often dispersed on a
porous support material to provide increased surface area. Ceria-
supported alumina was prepared by impregnation of alumina
aerogel pieces with a solution of ceric ammonium nitrate in
acetone after which the acetone was displaced by carbon dioxide
and the aerogel was dried under supercritical conditions and
then calcined at 723 K in air for 4 hours. Aerogels with ceria
loadings, for example, 20 weight % cerium retained higher surface areas than pure ceria aerogels (Figure 9.5). Ceria-doped
alumina aerogels containing 1 weight % palladium were more

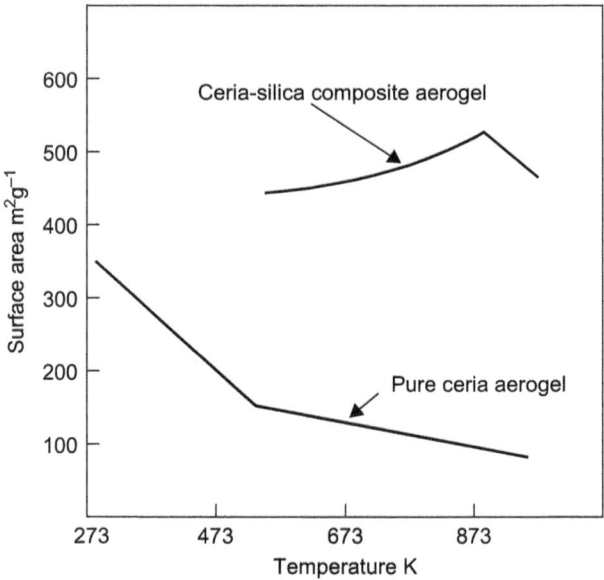

Figure 9.5   Surface area *versus* calcination temperature for silica aerogel containing 20 weight% Ce and pure ceria aerogels.[43]

active catalysts for the water–gas shift reaction (Equation 9.2) than pure ceria aerogels.

$$H_2O + CO \rightarrow H_2 + CO_2 \tag{9.2}$$

Another aerogel-based catalyst support was aluminium magnesium silicate derived from ion-exchange on the clay mineral montmorillonite and was active when containing the catalyst material for the addition polymerisation of monomers to polymers, for example, propylene polymerisation.[44] Aerogel-based catalyst supports containing ruthenium were active for synthesis of hydrocarbons from hydrogen and carbon monoxide in the Fischer-Tropsch reaction.[45] Other examples of aerogel supports in heterogeneous catalysis include a mixed iron/molybdenum catalyst dispersed on an alumina aerogel for the vapour phase deposition of single walled carbon nanotubes,[46] a hydrogenation catalyst based on a transition metal such as palladium dispersed on a silica aerogel and a mixed silica-titanium dioxide aerogel catalyst useful in the liquid phase process for the epoxidation of hydrocarbons with alkyl and aryl hydroperoxides.[47,48]

Applications for aerogels extend to space exploration.[49,50] They were used by NASA in the 1980s as a possible trap for capturing micrometeoroids during space flights and in the Mars Pathfinder probe to survey the atmosphere of Mars.

9.4 SUMMARY

Aerogels were discovered in 1932 and since then their properties, in particular their high surface area, small pore size, transparency, low thermal conductivity and high pore volume have attracted attention for diverse applications. These applications include thermal insulation, catalyst supports, anti-reflectance coatings, composite materials such as reinforced plastics and moulds for metal casting. The range of aerogels has expanded over the years from silica to organic aerogels, carbon aerogels, graphene and carbon nanotube aerogels and also cellulose-based aerogels.

REFERENCES

1. K. L. Yeung, N. Yao and S. Cao, Titania-silica aerogel monolith with ordered mesoporosity and preparation thereof, *United States Patent*, 8 222 302, 2012.
2. R. Defay, I. Prigogine, A. Bellemans and D. A. Everett, *Surface tension and adsorption*, Longman, London, 1966.
3. J. Zarzycki, Monolithic xero- and aerogels for gel-glass processes, in *Ultrastructure processing of ceramics, glasses and composites*, ed. L. L. Hench and D. R. Ulrich, John Wiley & Sons, New York, 1984, 27–42.
4. D. Segal, *Chemical synthesis of advanced ceramic materials*, Cambridge University Press, 1989.
5. F. J. Bonner, G. Kordas and D. L. Kinser, Sol-gel glasses by non-aqueous processes, *J. Non-Cryst. Solids*, 1985, **71**, 361–371.
6. W. A. W. I. Thielemans and R. Davies, Cellulose nanoparticle aerogels, hydrogels and organogels, *United States Patent Application*, 2013/0018112, 2013.
7. S. S. Kistler, Coherent expanded aerogels, *J. Phys. Chem.*, 1932, **36**, 52–64.
8. R. de Vos and G. L. J. G. Biesmans, Organic aerogels, *European Patent*, EP 0 710 262 B, 2000.

9. G. L. Biesmans and A. Mertens, Organic aerogels, *United States Patent*, 5 942 553, 1999.

10. G. A. Fox and T. M. Tillotson, Thio-, amine-, nitro- and macrocyclic containing organic aerogels & xerogels, *United States Patent*, 6 924 322, 2005.

11. S.-H. Park, M.-D. Cho, K.-H. Kim and S.-W. Hwang, Organic aerogel composition for forming the same, and method of preparing the same, *United States Patent*, 8 119 700, 2012.

12. J. K. Lee and G. L. Gould, Polyolefin-based aerogels, *United States Patent*, 7 691 911, 2010.

13. S. T. Mayer, J. L. Kaschmitter and R. W. Pekala, Carbon aerogel electrodes for direct energy conversion, *International Patent Application*, WO 95/20246, 1995.

14. C. Erkey and H. S. Hara, Aerogel and metallic compositions, *United States Patent*, 7 378 450, 2008.

15. A. F. Gross, J. J. Vajo, R. W. Cumberland, P. Liu and T. T. Salguero, Metal filled porous carbon, *United States Patent*, 7 910 199, 2011.

16. J. F. Cooper, T. M. Tillotson and L. W. Hrubesh, *United States Patent*, 7 410 718, 2008.

17. R. M. Hill, E. R. Fought and P. J. Biltoft, High specific area aerogel cryoadsorber for vacuum pumping applications, *United States Patent*, 6 122 920, 2000.

18. P. R. Coronado, Method for making monolithic metal oxide aerogels, *United States Patent*, 5 958 363, 1999.

19. B. Ziegler, N. Mronga, F. Teich and G. Herrmann, Hydrophobic silica aerogels, *United States Patent*, 5 738 801, 1998.

20. F. Schwertfeger, Method for the preparation of organically modified aerogels, *United States Patent*, 6 156 386, 2000.

21. H. H. Yokogawa, M. Y. Yokoyama, K. A. Takahama and Y. H. Uegaki, Process for forming a hydrophobic aerogel, *European Patent*, EP 0 653 377 B, 2004.

22. L. W. Hrubesh, P. R. Coronado and J. P. Dow, Method for removing organic liquids from aqueous solutions and mixtures, *United States Patent*, 6 709 600, 2004.

23. P. R. Coronado, J. F. Poco and L. W. Hrubesh, Super-hydrophobic fluorine containing aerogels, *United States Patent*, 7 211 605, 2007.

24. N. Leventis, J. C. Johnston, M. A. Kuczmarski and M. A. B. Meador, Surface modified aerogel monoliths, *United States Patent*, 8 394 492, 2013.
25. D. J. Kissel and C. J. Brinker, Durable polymer-aerogel based superhydrophobic coatings, a composite material, *United States Patent*, 8 663 742, 2014.
26. S. Giray, S. Kizilel and C. Erkey, Hydrophobic and hydrophilic aerogels encapsulated with PEG hydrogel via surface initiated photopolymerization, *United States Patent Application*, 2014/0065229, 2014.
27. T.-W. Kuo, LED package, *United States Patent Application*, 2012/0112223, 2012.
28. R.-M. Jansen, B. Kessler, J. Wonner and A. Zimmermann, Method for the subcritical drying of aerogels, *United States Patent*, 5 811 031, 1998.
29. R. Forbert, A. Zimmerman, D. M. Smith and W. Ackerman, Process for sub-critically drying aerogels, *United States Patent*, 6 131 305, 2000.
30. B. Ziegler and T. Gerber, Method of producing inorganic aerogels under subcritical conditions, *United States Patent*, 6 017 505, 2000.
31. M. Suzuki, T. Shiokawa, K. Yamada, H. Nakayama, H. Yamaguchi and A. Maruta, Production methods of silica aerogel film, anti-reflection coating and optical element, *United States Patent*, 7 931 940, 2011.
32. H. Nakayama, K. Yamada, Y. Sakai and M. Yamada, Method for producing silica aerogel coating, *United States Patent*, 8 029 871, 2011.
33. J.-G. Yeo, E. Lee, C.-H. Cho, H.-S. Park, N.-J. Park, N.-J. Jeong, C.-K. Hong and D.-K. Kim, Method for producing sheets including fibrous aerogel, *United States Patent*, 8 647 557, 2014.
34. K. Tsubaki, T. Kamae, H. Yokogawa, M. Yokoyama and K. Sonada, Optical fiber, *United States Patent*, 5 790 742, 1998.
35. M. Shaffer, A. G. Gallastegui, A. Asiri and S. Althabaiti, Cross-linked carbon nanotube networks, *United States Patent Application*, 2014/0012034, 2014.
36. A. Penicard, Aerogels of carbon nanotubes, *United States Patent Application*, 2011/0124790, 2011.

37. M. G. Schwab, K. Mullen, X. Feng and Z.-S. Wu, Aerogel based on doped graphene, *International Patent Application*, 2013/132388, 2013.
38. L. Ratke and S. Bruck, Production of aerogels containing fillers, *United States Patent*, 7 812 059, 2010.
39. M. K. Williams, T. M. Smith, J. E. Fesmire, L. B. Roberson and L. M. Clayton, Aerogel/polymer composite materials, *United States Patent*, 7 790 787, 2010.
40. J. Ryu, Flexible aerogel superinsulation and its manufacture, *United States Patent*, 6 068 882, 2000.
41. L. W. Hrubesh, J. F. Poco and P. R. Coronado, Fibers and fabrics with insulating, water-proofing, and flame-resistant properties, *United States Patent Application*, 2004/0142168, 2004.
42. M.-D. Cho, S.-H. Park, K.-H. Kim and S.-W. Hwang, Aerogel-foam composites, *United States Patent Application*, 2011/0311802, 2011.
43. E. M. Eyring, R. D. Ernst, G. C. Turpin and B. C. Dunn, Composite ceria-coated aerogels and methods of making the same, *United States Patent*, 8 435 918, 2013.
44. T. Sun, D. R. Wilson and J. M. Garces, Ion exchanged aluminium-magnesium silicate or fluorinated magnesium silicate aerogels and catalyst supports therefrom, *European Patent*, 1 066 110 B, 2003.
45. L. E. Manzer and K. Kourtakis, Fischer-Tropsch processes using xerogels and aerogel catalysts by destabilising aqueous colloids, *United States Patent*, 6 353 035, 2002.
46. J. Liu, High yield vapour phase deposition method for large scale single walled carbon nanotube preparation, *United States Patent Application*, 2003/0012722, 2003.
47. B. Heinrichs, J.-P. Pirard and R. Pirard, Transition metal aerogel-supported catalyst, *United States Patent*, 6 307 116, 2001.
48. A. Baiker, D. Dutoit and R. Hutter, Heterogeneous catalysts, *United States Patent*, 5 935 895, 1999.
49. S. A. Steiner III, Method for the formation of aerogel precursor using rapid gelation two-step catalysis, *United States Patent*, 6 764 667, 2004.
50. L. Reade, Full of air, *Chem. Ind.*, 2013, 77(2), 32–35.

Ionic Liquids

10.1 INTRODUCTION

Nowadays, there is a worldwide quest to identify technologies that can reduce carbon dioxide emissions into the environment for limiting the effects of global warming and to utilise renewable energy sources such as solar, hydroelectric, wind, tidal and geothermal. In addition, there are efforts to enhance recycling and reduce the generation of waste products in industrial processes and to reduce pollution. Solvents are a staple product of the chemical industry and there is increasing restriction through legislation on their use, recovery and disposal. Thus there is a demand for green chemical processes. Ionic liquids are a class of materials that are probably unknown to the general public and not widely known about within scientific circles. Ionic liquids have not been described as 'wonder materials', a phrase attributed to graphene, and they have escaped the widespread publicity associated with three-dimensional printing. This lack of publicity is unfortunate because, as an enabling technology, ionic liquids are finding use in an increasingly diverse range of applications. There are excellent reviews of ionic liquids but these rarely refer to the patent literature.[1] In this chapter an overview of ionic liquids is given, based mainly on developments

Exploring Materials through Patent Information
By David Segal
© David Segal, 2015
Published by the Royal Society of Chemistry, www.rsc.org

in the patent literature with a view to complementing reviews based on conventional journal articles.

10.2   IONIC LIQUIDS – DEFINITION AND NOMENCLATURE

Ionic liquids are liquids composed entirely of cations and anions derived from compounds that have high melting points or low melting points and, for the latter, at or below room temperature. Ionic liquids that have melting points below around 303 K are referred to as room temperature ionic liquids.[2] Usually the melting point is below 373 K and ionic liquids consist solely of ions, usually a bulky organic cation and an associated anion. In room temperature ionic liquids, the structures of the cation and anion prevent the formation of an ordered crystalline structure and the salt is therefore liquid at room temperature. The delocalised charge on one of the ions prevents the formation of a stable crystal lattice. Ionic liquids are liquid at temperatures below the individual melting points of the components and more than one species of cation and anion can be present in an ionic liquid. Alternative names have been used to describe ionic liquids, such as liquid electrolytes, ionic melts, ionic fluids, fused salts, liquid salts or ionic glasses. The variety of cations and anions allows the physical properties of ionic liquids, for example, viscosity, density, solubility in specific liquids to be tailored. Thus, their properties are tuneable. Ionic liquids containing hydrophobic anions such as hexafluorophosphates have very low solubilities in water, while ionic liquids containing hydrophilic anions such as chloride or acetate are completely miscible with water.[3] Ionic liquids that have a polymerisable group have been referred to as polymerisable ionic liquids. Conventional solvents such as acetone and dichloromethane are volatile organic compounds which contaminate soil and groundwater and add to air pollution generally. Ionic liquids have very low vapour pressures and represent a highly polar medium because of the charged species in them and are often viewed as environmentally friendly or 'green' alternatives to conventional organic solvents.

Examples of the range of cations and anions found in ionic liquids are shown in Table 10.1.[3,4]

Table 10.1 Examples of anions and cations in ionic liquids.[3,4]

Anion	Cation
$[AlCl_4]^-$	Pyridinium
$[Al_2Cl_7]^-$	Pyridazinium
NO_3^-	Pyrimidinium
BF_4^-	Pyrazinium
PF_6^-	Imidazolium
Alkyl- and arylsulphonates	Pyrazolium
Phosphates	Oxazolium
Imides	Triazolium
Methides	Thiazolium
Carboxylates	Piperidinium
	Pyrrolidinium
	Quinolinium
	Isoquinolium

Table 10.2 Nomenclature for ionic liquids.[4]

Ionic liquid	Abbreviation
1-butyl 3-methylimidazolium acetate	[BMIm]OAc
1-allyl-3-methylimidazolium chloride	[AMIm]Cl
1-ethyl-3-ethylimidazolium formate	[EMIm]OF

The names of ionic liquids are abbreviated according to a convention in which alkyl cations are often named by the first letters of the alkyl substituents and the cation that are given within a set of brackets and examples for the use of this nomenclature are given in Table 10.2.[3]

The structures of representative cations in ionic liquids are shown schematically in Figure 10.1 in which the groups R^1 to R^{10} can be H, halogen, OH, NH_2, SH or alky or aryl group.[5]

The sensitivity of the physical properties of ionic liquids on cation and anion type is highlighted in Table 10.3.[6]

10.3 EARLY DEVELOPMENT OF IONIC LIQUIDS

The development of room temperature ionic liquids dates back to the 1940s when a conducting bath of an electrolyte for the electrodeposition of aluminium was obtained from mixtures of anhydrous $AlCl_3$ and n-alkylpyridinium halides, for example,

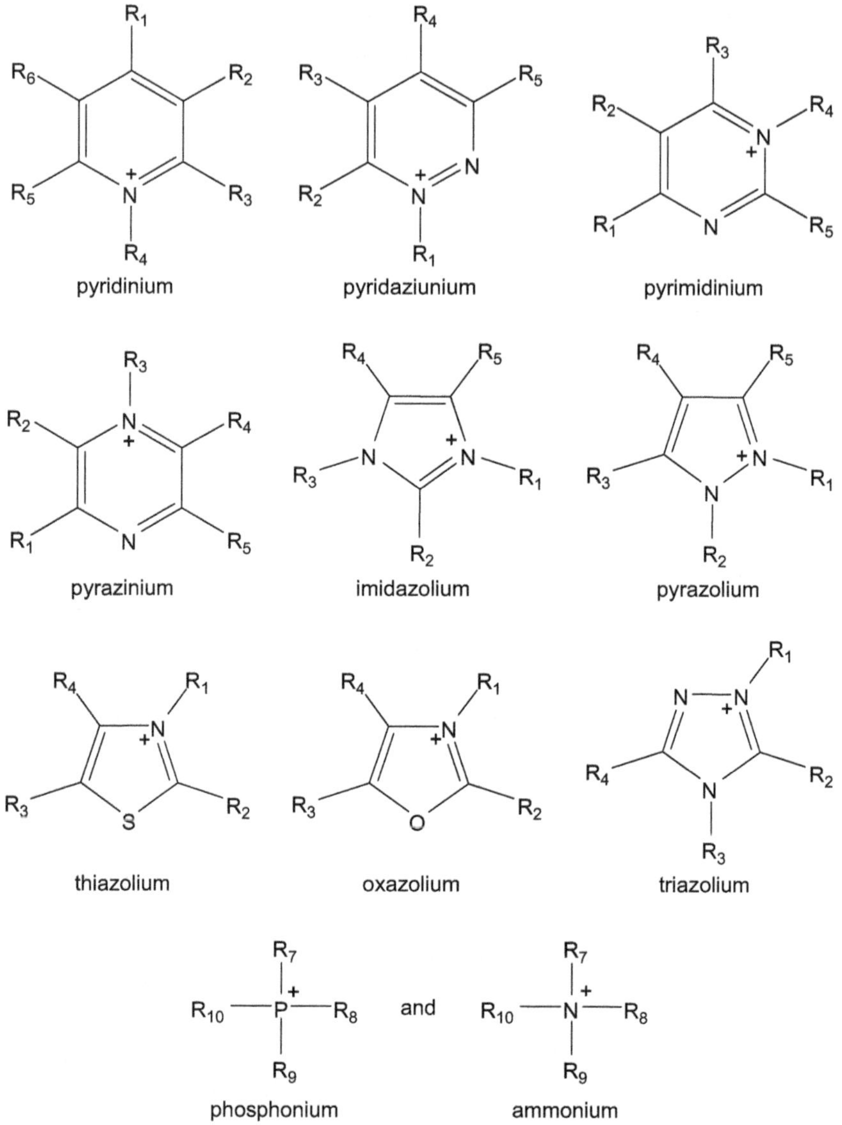

Figure 10.1   Examples of cations for ionic liquids.[5]

ethyl pyridinium bromide using a 2 : 1 mole ratio of aluminium chloride to ethyl pyridinium bromide.[7–9] This ratio prevented reduction of pyridinium cations at the cathode. The anode in the bath is aluminium and the cathode is the article to be plated.

Table 10.3 Effect of cation and anion on physical properties of ionic liquids.[6]

Ionic liquid	Melting point K	Viscosity mm s^{-1}	Density g cm^{-3}
1-ethyl-3-methyl imidazolium tetrafluoroborate	284	34	1.24
1-butyl-1-methylpyrrolidinium bis(trifluoromethyl sulphonyl)imide	223	71	1.4
1-butyl-3-methyl imidazolium hexafluorophosphate	284	312	1.36
1-butyl-3-methyl imidazolium tetraborate	202	233	1.21

However, a drawback to this system was the liberation of corrosive hydrogen chloride on exposure of $AlCl_3$ to water vapour. Electrolytes for use in electrochemical cells were obtained by heating a metal halide with what was referred to as a hydrocarbyl-saturated onium salt, for example, by heating $AlCl_3$ with trimethylphenylammonium chloride at 318 K in a 2 : 1 mole ratio.[10] The freezing point of the ionic liquid was 198 K compared to the melting points for trimethylphenylammonium chloride and $AlCl_3$ of 510 K and 456 K, respectively. Room temperature ionic liquids that had applications as catalysts, solvents and as electrolytes in batteries were prepared from mixtures of a metal halide such as aluminium chloride and an alkyl-containing amine hydrohalide salt, for example, trimethylamine hydrochloride or dibutylamine hydrochloride.[11] Analysis of the ionic liquid produced from trimethylamine hydrochloride indicated a formula $(CH_3)_3NHAl_2Cl_7$ and the presence of the polynuclear heptachloroaluminate anion $Al_2Cl_7^-$. The ionic liquids were active alkylation catalysts for conversion of, for example, benzene to cumene or dodecyl benzene.

Early applications for ionic liquids extend beyond electroplating. For example, an ionic liquid has been used to prevent build-up of electrostatic charge and the associated risk of explosion during transport of powdered sulphur by spraying an aqueous solution of the ionic liquid onto the powder surface.[12] Examples of suitable ionic liquids are cationic quaternary ammonium salt surfactants including stearyl-dimethyl-benzyl-ammonium chloride. Effective treatment of static charges required a liquid with both a low electrical resistivity and an ability to reduce the surface tension at the sulphur-liquid interface.

Interest in ionic liquids has continued to increase in recent years due in part to the requirement for 'green' solvents and utilisation of renewable energy sources and this is illustrated by the following examples from the patent literature.

10.4 APPLICATIONS FOR IONIC LIQUIDS

10.4.1 Hydrophobic Ionic Liquids

A disadvantage to the use of room temperature ionic liquids derived from $AlCl_3$ and n-alkylpyridinium halides is their hygroscopic nature and the release of hydrogen chloride on absorption of water vapour.[7–9] Hydrophobic ionic liquids do not absorb water vapour from the atmosphere and examples of such compounds are 1-ethyl-3-methylimidazolium perfluoro-1, 1-dimethylpropyl alkoxide and perfluoro-1-ethyl-3-methylimidazolium imide.[13] A feature of these hydrophobic liquids was the requirement for use of an anion that is a non-Lewis acid–containing polyatomic anion having a van der Waals volume greater than 0.1 nm³. The anions $(CF_3SO_2)_2N^-$ and $(CF_3SO_2)_3C^-$ have van der Waals volumes of 0.143 nm³ and 0.206 nm³, respectively, and are appropriate for hydrophobic ionic liquids. In contrast to these values the anions $CF_3SO_3^-$ and BF_4^- have van der Waals volumes of 0.08 nm³ and 0.048 nm³, respectively, and are soluble in water.

In general, increasing the length of the alkyl chain tends to decrease water solubility by increasing the hydrophobicity of the cation, for example, pyridinium or imidazolium.

Table 10.4 shows the melting points for a selection of tetra-alkyl ammonium halides.[14] Hydrophobic room temperature

Table 10.4 Melting points for tetra-alkyl ammonium halides.[14]

Compound	Melting point K
Tetrabutyl ammonium bromide	373–376
Tetrabutyl ammonium hydrogen sulphate	442–445
Tetrabutyl ammonium tetrahydridoborate	397–402
Tetrabutyl ammonium iodide	416–419
Tetrapentyl ammonium iodide	408–410
Tetrahexyl ammonium iodide	394–397
Tetra-octyl ammonium bromide	368–371

ionic liquids based on tetra-alkyl ammonium halides were obtained when one of the substituent groups was a benzyl group or alkyl group containing 1 to 4 carbon atoms and the other three substituent groups were all different and contained 8 to 10 carbon atoms. The anion was a halide or alkyl sulphate. Ionic liquids based on tetra-alkyl phosphonium salts were also obtained. Applications for the ionic liquids include solvents, phase transfer catalysts and liquid media for batteries.

Another class of hydrophobic ionic liquids is based on alkylpyridinium dicyanamides.[15] While n-butylpyridinium dicyanamide and n-butyl-3-methylpyridinium dicyanamide are water-miscible, n-octylpyridinium dicyanamide and n-octyl-3-methylpyridinium dicyanamide are water-immiscible. Applications for these liquids are as polar solvents.

10.4.2 Use of Ionic Liquids for Treatment of Biomass and Biofuel Production

Lignocellulose, often referred to as biomass, has three main components;[16] 30–40 weight % cellulose, 20–30 weight % hemicellulose and 5–30 weight % lignin. It is a major structural component of plants and is found in waste materials including sawdust, wood chips, straw and bagasse (sugar cane residue). Cellulose is the most abundant biorenewable material on earth and consists of linear polymeric chains formed by repeated connection of β-D-glucose building blocks through a 1-4 glycoside linkage.[17] Cellulose is crystalline in nature in which the polymer chains are held together by hydrogen bonding and van der Waals forces. The hydrogen bonded structures are insoluble in water and common organic solvents. The structure of cellulose and β-D-glucose is shown in Figure 10.2.[17]

Hydrolysis of cellulose (saccharification) generates monosaccharide, disaccharide and oligosaccharide products and glucose is the main hydrolysis product. Cellulose hydrolysis is catalysed by mineral acids or enzymes, in particular by cellulases. Glucose is an important material for fermentation to ethyl alcohol so that saccharification is relevant to the production of biofuel but the crystalline matrix of cellulose is difficult to penetrate by water molecules and catalysts. In a process for obtaining water-soluble cellulose hydrolysis products samples

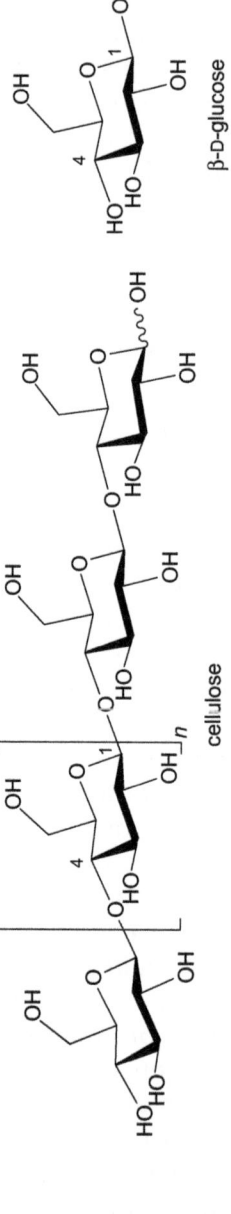

Figure 10.2   Schematic diagram of the structure of cellulose and β-D-glucose.[17]

of cellulose were dissolved in an ionic liquid such as 1-n-butyl-3-methylimidazolium chloride (BMIMCl) or 1-allyl-3-methylimidazolium chloride (AMIMCl) at around 406 K.[16] Cellulose was precipitated by addition of an anti-solvent such as water, methanol or ethanol and was amorphous compared with untreated cellulose that was highly crystalline. The precipitated cellulose could be enzymatically hydrolysed to sugars using cellulose. In another application of ionic liquids cellulose was partially dissolved in, for example, tributylmethyl ammonium chloride at 395 K in the presence of an acid, trifluoroacetic acid, CF_3COOH to aid hydrolysis.[17] This method produced high yields of water-soluble products having glucose end groups, hence 8.7 weight % after 150 minutes. Halide-based ionic liquids such as 1-allyl-3-methylimidazolium chloride have also been used in a process for the production of cellulose esters from cellulose.[4]

Bio-diesel is defined as the mono-alkyl esters of long chain fatty acids derived from vegetable oils or animal fat.[18] It is made either by transesterification of animal fat and separation of glycerine from the fatty acid methyl ester or by esterification of vegetable oils followed by separation of the water by-product from the fatty acid methyl ester. In an application for ionic liquids, bio-diesel was produced by esterification of fatty acids in vegetable oil by using an ionic liquid that acts as both a solvent and catalyst, for example, ionic liquids based on the pyrazolium cation. Another source of bio-diesel is from algae in the oceans.[19] A general approach to harvest the products contained in algae cells, namely lipids, hydrocarbons and carbohydrates involves contacting dry algae with an ionic liquid, 1-butyl-3-methylimidazolium chloride that resulted in cell lysis and dissolution, thus releasing the contents of the cells that can be separated and processed further for bio-fuels.

10.4.3 Ionic Liquids as Solvents and Reactive Solvents

Nuclear fuel in the form of oxide pellets are clad in Zircalloy rods and inserted into the core of a nuclear reactor. After extensive irradiation the rods are removed, stored under water for a period of time to cool down, after which the rods are chopped up and the pellets removed for reprocessing. Traditional reprocessing involves initial dissolution of radioactive pellets in nitric acid but ionic liquids have also been considered in an alternative approach.[20] Here, an oxide, for example, UO_2 was dissolved in a mixture of 1-butylpyridinium nitrate and a nitronium tetrafluoroborate $[NO][BF_4]$ at 363 K. The nitronium ion NO^+ enhances the oxidising power of the ionic liquid that oxidises UO_2 to the UO_2^{2+} in solution, thus a change in oxidation state from IV to VI. A key feature of the process is that uranium is more soluble in the VI state than the IV state and this is important when handling radioactive materials, as the amount of solvent required is reduced. Sulphur is an important raw material in the chemical industry but there are few solvents that offer both good solubility and are environmentally safe to use. Carbon disulphide (CS_2) is a well-known solvent for sulphur but it is toxic, volatile, flammable and has a low boiling point. Ionic liquids have been considered as solvents for sulphur and also phosphorus, selenium and tellurium.[21] Suitable anions for the ionic liquid had high polarisability and low electronegativity; so-called soft anions combined with, for example, the butyl methyl imidazolium cation. Examples of suitable anions were $S_2CSC_4H_9^-$, $S_2CN(CH_2)_4^-$ and $S_2COC_2H_5^-$. Ionic liquids offer an alternative approach for purification of phosphorus, selenium, tellurium and sulphur. A further application of ionic liquids is as solvents for chemical vapour deposition.[22] In the latter process a non-volatile solid coating is deposited onto a substrate such as a semiconductor wafer using vapour phase reactants. In some cases a solid precursor is dissolved in a solvent and the solution is then vapourised, but the solvent and precursor can decompose and decomposition products block delivery lines in the deposition chamber. When an ionic liquid is used as solvent, for example, 1-ethyl-3-methyl-1H-imidazolium tetrafluoroborate, precursors for a thin oxide coating on a substrate are dissolved in the liquid in a gas bubbler and the precursor is carried over in the vapour phase to the substrate. As the ionic liquid has a low vapour pressure it is not carried over in the vapour phase.

Table 10.5 shows the freezing point of ionic liquids formed between choline chloride, $(CH_3)_3(CH_2)_2OHNCl$ and a range of amides, carboxylic acids and alcohols.[23] Melts could only be produced when the organic compound formed a hydrogen bond with the halide anion. Thus ionic liquids were not produced for mixtures of choline chloride and cyclohexane, pyridine or ethyl

Table 10.5 Freezing point for ionic liquids containing choline chloride and an organic compound.[23]

Organic compound	Freezing point K
Urea NH_2CONH_2	285
Acetamide CH_3CONH_2	324
Thiourea NH_2CSNH_2	342
Benzoic acid C_6H_5COOH	368
Phenol C_6H_5OH	243

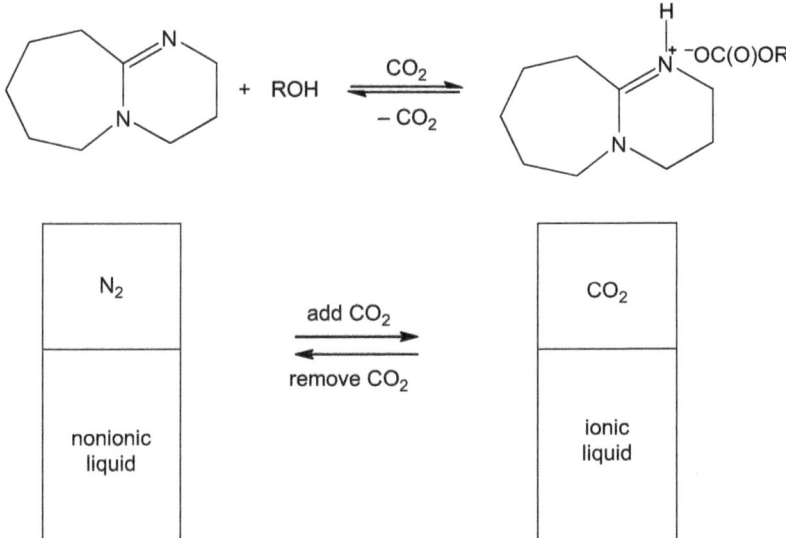

Figure 10.3 Schematic diagram of a switchable ionic liquid.[24]

alcohol as examples. These ionic liquids acted as solvents for precious metal oxides such as platinum and palladium in spent catalysts recovered from catalytic converters, especially when the organic compound was oxalic acid.

Solvents that can reversibly form ionic liquids by exposure to a chemical trigger have been developed and an example is shown schematically in Figure 10.3.[24] A mixture of an alcohol such as 1-hexanol and 1,8-diazabicyclo-[5,4,0]-undec-7-ene (DBU) is present under a blanket of nitrogen but an ionic liquid forms when carbon dioxide is bubbled through the mixture producing an amidinium alkyl carbonate. These switchable solvents have potential applications as CO_2 sensors but can also dramatically change solvent properties so that a species in solution can be precipitated out of solution.

Nitrous oxide, N_2O is a by-product of various chemical processes, for example, in the nitric acid oxidation of cyclohexanone in the synthesis of adipic acid that is used for manufacture of nylon 6,6.[25] However, N_2O is a powerful greenhouse gas and it is desirable to cut emissions of this gas into the atmosphere. Nitrous oxide can be absorbed into an ionic liquid, for example, 1-n-butyl-3-methylimidazolium[Ga_2Cl_7] thus potentially reducing emissions to atmosphere and used in chemical conversions such as the oxidation of cyclododecene to cyclododecanone. Further examples of reactive ionic liquids are those that are susceptible to electrochemical reduction with applications as electrolytes in batteries.[26]

10.4.4 Ionic Liquids in Separation Processes

Industrial chemical processes often result in products either in the gaseous phase or in solution that need to be separated from each other. Ionic liquids show promise in facilitating separation processes, as illustrated by the following examples. Oil and gas fields are often contaminated with mercury and apart from its toxicity, mercury can form amalgams with aluminium components of processing equipment, leading to equipment failure. Conventional methods for removing mercury include scrubbing using fixed beds of sulphur or sulphur-containing materials such as transition metal sulphides. Ionic liquids containing a cation with an oxidation state greater than two show much promise in removing mercury vapour from hydrocarbon feeds.[27] For example, a copper(II)-containing 1-butyl-3-methylimidazolium chloride ionic liquid that was supported on activated carbon or porous silica supports. A possible role for the ionic liquid was as a mild oxidising agent to oxidise elemental mercury species to soluble ionic species.

Tetrafluoromethane, CF_4, is used widely as a plasma etchant for fabrication of semiconductor devices in order to lay down electrical pathways on the surface of an integrated circuit wafer.[28] In addition, nitrogen trifluoride, NF_3, is used in plasma form as a cleaning gas to remove surface contamination from the internal surfaces of semiconductor fabrication equipment. Exhaust gases from the manufacturing processes include unreacted NF_3 and CF_4 and fluorinated reaction products such as hydrogen fluoride. There are pressing environmental reasons as well as economic reasons because of the cost of reactants for capturing the highly toxic perfluorinated reaction products. Mixtures of NF_3 and CF_4 have been separated by use of an ionic liquid. For example, NF_3 has a higher solubility in 1-butyl-3-methylimidazolium hexafluorophosphate than CF_4 enabling components in a mixture to be separated, after which CF_4 is recovered by distillation. Mixtures of hydrogen chloride and phosgene ($COCl_2$) are obtained in many industrial processes, for example, in the preparation of isocyanates and polycarbonates and have to be separated, and this has been achieved by use of ionic liquids.[29] For example, ethylmethylimidazolium chloride preferentially absorbs hydrogen chloride from a mixture of HCl and $COCl_2$ at low temperatures around 313 K. Dissolved hydrogen chloride is then purged from the ionic liquid that is exposed to the gaseous mixture, now depleted in HCl, to repeat the recovery cycle.

Pure sugars are required in many applications.[30] For example, high fructose corn syrup is extensively used in the food industry while pure glucose is used for medical purposes and in the manufacture of pharmaceuticals. Fructose and glucose occur together in crops and while chromatography is a common method for commercial separations, other separation techniques are under evaluation. Thus the ionic liquid 1,3-dimethylimidazolium dimethylphosphate preferentially dissolves glucose in a mixture of glucose and fructose at ambient temperatures, leaving behind a fructose precipitate. Exposure of the residual mixed sugar solution to another ionic liquid 1-ethyl-3-methylimidazolium ethylsulphate resulted in preferential dissolution of fructose and precipitation of glucose. Use of these cycles facilitated sugar separations, typically requiring 100 g of ionic liquid to recover 57 g of glucose. Ionic liquids have shown promise in the extraction of gold ions, $AuCl_4^-$ from aqueous solution.[31] Examples of ionic liquid used for the extraction process were ethyl-3-methylimidazolium bis(trifluoromethanesulphonyl)imide and 1-methyl-1-propylpyrrolidinium bis(trifluoromethanesulphonyl)imide.

Amidium-based ionic liquids are formed by reaction of an amide such as *N,N*-dibutylformamide and *N,N*-diethylbenzamide with an organic acid, for example, chloroacetic acid and propionic acid.[32,33] Traditional methods for removing carbon dioxide from exhaust gases in industrial plants, or in natural gas, is by chemical reaction with aqueous amine-based solutions, after which CO_2 is stripped from the solution at high temperature in an energy-intensive process to regenerate the solution. Amidium-based ionic liquids, for example, *N,N*-dimethylformamidium trifluoroacetate show promise for absorption of CO_2 at 313 K and pressures up to 2 M Pa. Regeneration of the ionic liquid was accomplished by releasing the reactor pressure, a potentially less energy intensive process than regeneration of amine-based solutions.

Uses for ionic liquids are not restricted to the laboratory-scale, and the Basil process (biphasic acid scavenging utilising ionic liquids) is an industrial process that uses ionic liquids.[34] Acids produced in chemical reactions have been removed by addition of amines, producing crystalline salts that are removed by filtration. However, this procedure can block up equipment. In the Basil process acids are removed from reaction solvents by interaction with a base in the form of an ionic liquid and the resulting liquid, rather than solid, crystals can be separated from the reaction product by decanting.

10.4.5 Catalytic Reactions in Ionic Liquids

Alkylation of aromatic compounds has traditionally been carried out by reaction of an alkyl halide with an aromatic hydrocarbon in the presence of a Lewis acid catalyst such as hydrofluoric acid, boron trifluoride and zeolites, as well as combinations of acidic catalysts such as aluminium chloride and alkyl aluminium halides.[35] These reactions require expensive equipment and strong acids. A simpler alkylation process was developed in which benzene was alkylated with ethylene in an autoclave at 2.4 M Pa and 383 K in the presence of an ionic liquid catalyst comprising 67 weight % aluminium chloride and 33 weight % 1-ethyl-3-methyl-imidazolium chloride. Ionic liquids have been developed as catalyst supports for the polymerisation of olefins and a representative reaction scheme for preparation of the support is shown in Figure 10.4.[36] A bisimine (Figure 10.4(a)) is reacted with lithium diisopropylamide and an alkyl halide to produce a halogenated bisimine (Figure 10.4(b)). The halogenated bisimine is then reacted with an ionic liquid precursor such as an

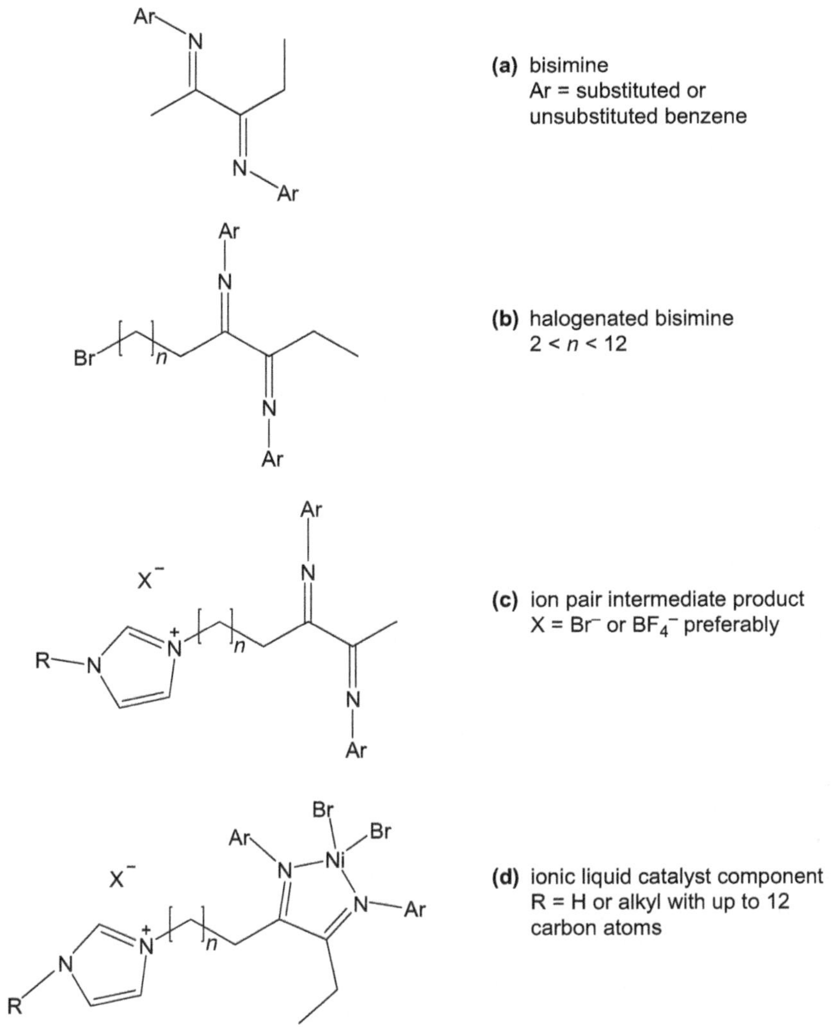

(a) bisimine
Ar = substituted or
unsubstituted benzene

(b) halogenated bisimine
$2 < n < 12$

(c) ion pair intermediate product
X = Br$^-$ or BF$_4$$^-$ preferably

(d) ionic liquid catalyst component
R = H or alkyl with up to 12
carbon atoms

Figure 10.4 Schematic diagram illustrating the preparation of an ionic liquid catalyst component.[36]

n-alkylimidazole to yield the intermediate product, an ion pair shown in Figure 10.4(c), for example, a bisimine imidazolium tetrafluoroborate. This intermediate is then reacted with a metallocene precursor containing nickel or palladium resulting in the ion pair shown in Figure 10.4(d). The catalyst system consists of a material of the type illustrated in Figure 10.4(d) and an activating agent, methylalumoxane. The catalyst was active for polymerisation of ethylene at 298 K and 0.4 M Pa.

Isopentane is produced in refineries during hydrocarbon upgrading processes such as hydrocracking and while it had been used as a component for gasoline, its high volatility has placed environmental limitations on its use as a blending component. The room temperature ionic liquid n-butyl pyridinium chloroaluminate, $C_5H_5NC_4H_9Al_2Cl_7$, is an alkylation catalyst, for example, for the reaction of ethylene with isopentane at 323 K and at pressures up to 24 M Pa.[37] Because hydrocarbons have a low solubility in ionic liquids the reaction is biphasic and takes place at the interface with the ionic liquid. The major reaction product, a C_7 alkane has potential as a blending component for hydrocarbon fuels. Ionic liquids containing a transition metal complex have been used as hydrosilylation catalysts.[38] For example, the ionic liquid 1-ethyl-3-methylimidazolium bistrifluoromethanesulphonylimide containing a solution of platinum tetrachloride could be used either as a solution or when applied as a coating to a porous support such as granular silica in what has been referred to as supported ionic liquid phase (SILP) catalyst technology.

Formic acid is an important intermediate in the production of pharmaceutical and agricultural products and its preparation has involved the use of metal catalysts and supercritical carbon dioxide.[39] A milder synthesis of formic acid involves the reaction between carbon dioxide and hydrogen in an ionic liquid without using supercritical conditions and in the absence of a metal catalyst; an example of a suitable ionic liquid for the synthesis is 1,3-di-n-propyl-2-methylimidazolium formate. Further examples of the use of ionic liquids in catalytic reactions include the production of ethylene glycol with quaternary ammonium or quaternary phosphonium ionic liquids and polyoxymethylene dimethyl ethers from methylal and trioxymethylene at temperatures in the range 368–403 K and reaction pressures between 0.8 and 4.0 M Pa.[40,41]

10.4.6   Ionic Liquids and Pharmaceuticals

Synthetic peptides have various applications, including in functional foods, nutrient compositions, livestock feeds and active ingredients for pharmaceutical products. A method for producing peptides in high yields by using ionic liquids has been described.[42] As an example, the amino acid benzyloxy-carbonyl aspartic acid was added to a solution of tetrabutyl phosphonium hydroxide, after which water was removed under vacuum that encouraged a dehydration-condensation reaction leading to formation of benzyloxy – carbonyl aspartic acid—tetrabutyl phosphonium (reactant I), a colourless transparent liquid. Separately, phenylalanine was reacted with methanol and sulphuric acid to form phenylalanine methyl ester—monomethyl sulphate (reactant II). Reactant II dissolved in the ionic liquid reactant I and after raising the temperature to 310 K and adjusting the pH to 6.0, the enzyme thermolysin was added to the mixture. Termination of the enzymatic reaction after 48 hours and analysis of the solution by high pressure liquid chromatography identified benzyloxy-carbonyl aspartic acid phenylalanine methyl ester.

Non-steroidal anti-inflammatory drugs (NSAID) such as diclofenac are often administered as an oral preparation for inflammatory diseases, but they are associated with side effects involving the digestive tract.[43,44] The use of ionic liquids to administer NSAIDs as an external preparation has been explored.[43] For example, an ionic liquid was formed between a carboxylic acid such as diclofenac and an amine, eperisone hydrochlorate and dissolved in diethyl sebacate, a carrier liquid used for external preparations. The ionic liquid has the potential for use in pharmaceutical compositions that are applied transdermally through the skin. Ionic liquids derived from a carboxylic acid having five to twenty carbon atoms and an amine having four to twelve carbon atoms for example isostearic acid and diisopropanolamine or lauric acid and triethanolamine as examples have been considered as solvents for administering drugs such as lidocaine transdermally through the skin.[44]

10.4.7 Ionic Liquids and Energetic Materials

The energetic material 2,4,6-trinitrotoluene (TNT) is a melt-castable explosive but its toxic vapour presents a hazard to anyone involved in its manufacture.[45] Ionic liquids based on triazolium salts have been evaluated as replacements for TNT in melt-castable explosives. The low vapour pressure of ionic

Table 10.6   Energetic materials derived from ionic liquids based on triazolium salts.[45]

Compound	Melting point K
TNT	354
4-amino-1,2,4-triazolium perchlorate (4-ATP)	357
4-amino-1,2,4-triazolium nitrate (4-ATN)	not available
1-H-1,2,4,-triazolium perchlorate (TP)	not available
1-amino-3-methyl-1,2,3-triazolium nitrate (1-AMTN)	361
1-methyl-4-amino-1,2,4-triazolium perchlorate (MATP)	358
70 weight % 4-ATP – 30 weight % ATN eutectic	339

liquids reduces the toxic risk that employees are exposed to during manufacture. Table 10.6 lists potential replacements for TNT and all of these materials had low melting points, particularly eutectic mixtures, which is a desirable property for melt-casting. Ionic liquids have been considered as energetic materials for propellants that are often based on hydrazine, a toxic and carcinogenic material. An oxygen balance is required between the fuel and oxidiser in the system for complete combustion of the propellant and in the case of ionic liquids, the fuel is represented by bulky cations and the oxidiser is represented by incorporation of energetic groups into the cation and oxidising anions.[46] Examples of ionic liquids as potential replacements for hydrazine-based propellants include 1-ethyl-3-methylimidazolium tetranitratoborate.

Peroxide-based explosives such as triacetone triperoxide (TATP) and hexamethylene triperoxide diamine (HMTD) can be produced from household chemicals and can be highly unstable.[47] Ionic liquids have been developed that dissolve peroxide-based explosives allowing safe handling of these materials. Examples of suitable ionic liquids include 1-octyl-3-methylimidazolium tetrafluoroborate and 1-ethyl 3-methylimidazolium bis(trifluoromethanesulphonimide).[47]

10.4.8   Ionic Liquids and Adhesives

Operations that are carried out in spacecraft involve fabrication and repair procedures using adhesives and composites.[48] As repairs are carried out in confined spaces it is important that there are no toxic gases produced from the adhesive while structural components fabricated from the adhesives must withstand extreme temperatures and hard vacuum. Ionic liquid epoxide monomers that can be cured to a polymeric cross-linked epoxy resin for use in harsh environments have been developed. As an example, 1,3-bis(glycidyl)imidazolium tri-fluoromethanesulphonimide monomer derived from the reaction of epichlorohydrin and imidazole was cured in the presence of an amine hardener, 1,3-bis(3-aminophenoxy) benzene at 393 K. The cured resin had a tenslle strength of 5.7 M Pa and adhesive strength of 4.0 M Pa at room temperature and a tensile strength of 4.5 M Pa at 77 K with a decomposition temperature higher than 523 K. Composites were prepared between monomer and graphite fabrics. Pressure-sensitive sheets containing an anti-static adhesive are used to prevent the build-up of electric charge on insulating components such as plastics and also to protect the surface of components from scratch marks.[49,50] Ionic liquids have been used as a component in the adhesives. For example, 1-butyl-3-methylpyridinium bis(trifluoromethanesulphonyl) imide in conjunction with an acryl-based polymer derived from a mixture of 2-ethylhexyl acrylate, ethoxy-diethylene glycol acrylate and 2-hydroxyethyl acrylate and a polymerisation initiator 2,2′-azobisisobutyronitrile.

10.5 SUMMARY

Ionic liquids are liquids composed entirely of cations and anions. While room temperature ionic liquids found initial application as electrolytes for electrodeposition of aluminium in the 1940s, interest in them has increased markedly in recent years, partly due to the growth in green chemical processes but also because of their versatility for a wide range of applications. The range of applications include: (i) solvents for selected dissolution for example dissolution of sulphur; (ii) the processing of vegetable oils to produce bio-diesel; (iii) separation of components in a mixture, for example, glucose from fructose; (iv) catalysts for a range of reactions including alkylation of hydrocarbons; (v) the production of pharmaceuticals; (vi) safer alternatives to conventional explosives; and (vii) adhesives. The low vapour property of ionic liquids is a very attractive property for their application. In addition, an industrial process that uses ionic liquids for removal of acidic by-products in chemical processes has been developed.

REFERENCES

1. F. Kerton and R. Marriott, *Alternative Solvents for Green Chemistry*, Royal Society of Chemistry, 2nd edn, 2013.
2. R. D. Rogers, J. Holbrey and H. Rodrugues, Process for removing metals from hydrocarbons, *United States Patent Application*, 2012/0121485, 2012.
3. C. M. Buchanan and N. L. Buchanan, Cellulose esters and their production in halogenated liquids, *United States Patent*, 8 273 872, 2012.
4. N. Ignatyev, U. Welz-Biermann, M. Finze, E. Bernhardt and H. Willner, Salts having alkoxytris(fluoroalkyl)borate anions, *United States Patent*, 8 143 452, 2012.
5. M. A. Harmer, C. P. Junk and J. Vickery, Ionic liquids, *United States Patent*, 7 834 192, 2010.
6. S. K. Chun, I.-S. Hwang, D. W. Lee and Y. K. Son, Printing composition and a printing method using the same, *United States Patent Application*, 2013/0059135, 2013.
7. T. P. Wier, Electrodeposition of aluminum, *United States Patent*, 2 446 350, 1948.
8. F. H. Hurley, Electrodeposition of aluminum, *United States Patent*, 2 446 331, 1948.
9. T. P. Wier and F. H. Hurley, Electrodeposition of aluminum, *United States Patent*, 2 446 349, 1948.
10. S. D. Jones and G. E. Blomgren, Low temperature molten composition, *United States Patent*, 4 764 440, 1988.
11. F. G. Sherif, L.-J. Shyu, C. P. M. Lacroix and A. G. Talma, Low temperature ionic liquids, *United States Patent*, 5 731 101, 1998.
12. R. Lacroux and J.-P. Zwilling, Process for suppressing electrostatic charges on sulphur, *United States Patent*, 3 282 728, 1966.
13. V. R. Koch, C. Nanjundiah and R. T. Carlin, Hydrophobic ionic liquids, *United States Patent*, 5 827 602, 1998.
14. H. Huesken, P. Birnbrich and G. Schenker, Compounds that are liquid at ambient temperature, *United States Patent Application*, 2010/0204074, 2010.

15. C. Taschler and G. Clausen, Alkyl pyridinium dicyanamides and method for the production thereof, *United States Patent Application*, 2012/0330023, 2012.
16. S. Varanasi, C. A. Schall and A. P. Dadi, Saccharifying cellulose, *United States Patent*, 7 674 608, 2010.
17. J. Holbrey, M. Fanselow, K. R. Seddon, L. Vanoye and A. Zheng, Conversion method, *United States Patent*, 8 574 368, 2013.
18. M. J. Earle, K. R. Seddon and N. V. Plechkova, Production of bio-diesel, *United States Patent Application*, 2009/0235574, 2009.
19. R. E. Teixeira, Process for the extraction of lipids from microalgae using ionic liquids, *United States Patent Application*, 2011/0217777, 2011.
20. M. Fields, G. V. Huston, K. R. Seddon and C. M. Gordon, Ionic liquids as solvents, *United States Patent*, 6 379 634, 2002.
21. M. J. Earle, E. Boros, K. R. Seddon, M. A. Gilea and J. S. Vyle, Elemental solvents, *United States Patent Application*, 2010/0178229, 2010.
22. S. Uhlenbrock, Chemical vapour deposition methods utilizing ionic liquids, *United States Patent*, 6 998 152, 2006.
23. A. P. Abbott, D. L. Davies, G. Capper, R. K. Rasheed and V. Tambyrajah, Ionic liquids and their use as solvents, *United States Patent*, 7 183 433, 2007.
24. P. G. Jessop, C. A. Eckert, C. L. Liotta and D. J. Heldebrant, Switchable solvents and methods of use thereof, *United States Patent Application*, 2013/0327989, 2013.
25. S. Aki, K. Whiston, T. Belhocine and K. R. Seddon, Nitrous oxide-containing ionic liquids as chemical reagents, *United States Patent Application*, 2013/0299738, 2013.
26. M. Schmidt, N. M. Ignatyev and W.-R. Pitner, Reactive ionic liquids, *United States Patent Application*, 2011/0045359, 2011.
27. M. Abai, M. P. Atkins, K. Y. Cheun, J. Holbrey, P. Nockemann, K. Seddon, G. Srinivasan and Y. Zou, Process for removing metals from hydrocarbons, *United States Patent Application*, 2014/0001100, 2014.
28. M. B. Shiflett and A. Yokozeki, Process for purifying perfluorinated products, *European Patent*, 2 021 276 B, 2013.
29. A. Woelfert, C. Knoesche, H.-J. Pallasch, M. Sesing, E. Stroefer, H.-M. Polka and M. Hellig, Method for separating

hydrogen chloride and phosgene, *United States Patent*, 7 659 430, 2010.

30. I. M. Al Nashef, M. H. Gaily, S. M. Al-Zahrani and A. E. Abasaeed, Method for separating fructose and glucose, *United States Patent*, 7 942 972, 2011.

31. J. Ralston, J. Zhou, C. I. Priest and R. V. Sedev, Extraction of gold, *International Patent Application*, WO 2012/079129, 2012.

32. S. J. Choi, J. Palgunadi, J. E. Kang, H. S. Kim and S. Y. Chung, Amidium-based ionic liquids for carbon dioxide absorption, *United States Patent*, 8 613 865, 2013.

33. S. J. Choi, J. Palgunadi, J. E. Kang, H. S. Kim and S. Y. Chung, Amidium-based ionic liquids for carbon dioxide absorption, *United States Patent*, 8 282 710, 2012.

34. K. Ekman, M. Carla, M. Sundell, V. Suhonen and J. Hautojarvi, Method for recovering metals from a spent catalyst composition, *United States Patent*, 8 282 705, 2012.

35. A. K. Abdul-Sada, M. P. Atkins, B. Ellis, P. K. G. Hodgson, M. L. M. Morgan and K. R. Seddon, Alkylation process, *United States Patent*, 5 994 602, 1999.

36. O. Lavastre, F. Bonnette and A. Razavi, Ionic liquid supports, *United States Patent Application*, 2007/0213485, 2007.

37. H. K. C. Timken, S. Elomari, S. Trumbull and R. Cleverdon, Integrated alkylation process using ionic liquid catalysts, *United States Patent*, 7 432 408, 2008.

38. A. Bauer, T. Frey, N. Hofmann, P. Schulz and P. Wasserscheid, Method for production of organosilicon compounds by hydrosilylation in ionic liquids, *United States Patent Application* 2010/0267979, 2010.

39. M. Nakahara, N. Matsubayashi and Y. Yasaka, Method for producing formic acid, *United States Patent*, 8 519 013, 2013.

40. S. Zhang, J. Sun, W. Cheng, J. Wang, J. Zhang, Z. Fu and X. Zhang, Process for producing ethylene glycol catalysed by ionic liquid, *United States Patent*, 8 658 842, 2014.

41. J. Chen, H. Song, C. Xia, X. Zhang and Z. Tang, Method for synthesizing polyoxymethylene dimethyl ethers by ionic liquid catalysis, *United States Patent*, 8 344 183 2013.

42. S. Furukawa, K. Hasegawa and I. Fuke, Method for producing peptide, *United States Patent Application*, 2013/0143262, 2013.

43. N. Hanma, Y. Miwa and H. Hamamoto, Salt of nonsteroidal anti-inflammatory drug and organic amine compound and use thereof, *United States Patent Application*, 2010/0029704, 2010.
44. T. Yamaguchi, K. Kawai, K. Yamanaka and N. Tatsumi, External preparation composition comprising fatty acid-based ionic liquid as active ingredient, *United States Patent Application*, 2014/0066471, 2014.
45. T. W. Hawkins, G. W. Drake and A. J. Brand, Energetic ionic liquids, *United States Patent*, 7 645 883, 2010.
46. K. O. Christie and G. W. Drake, Energetic ionic liquids, *United States Patent*, 7 771 549, 2010.
47. S. Waldvogel, C. Siering, D. Lubczyk, J. Loebau and A. Hahma, Odor samples of peroxidic explosives, *United States Patent*, 8 603 270, 2013.
48. M. S. Paley, R. S. Libb, R. N. Grugel and R. E. Boothe, Ionic liquid epoxy resins, *United States Patent*, 8 193 280, 2012.
49. T. Amano, N. Kobayashi and M. Ando, Pressure-sensitive adhesive composition, pressure-sensitive adhesive sheets and surface protecting film, *United States Patent*, 7 989 525, 2011.
50. N. Ukei, T. Amano and M. Ando, Adhesive composition, adhesive sheet and surface protective film, *European Patent Application*, 1 939 263 A, 2008.

CHAPTER 11

Flame Retardants

11.1 INTRODUCTION

Flame retardants are chemical additives that are incorporated into products such as plastics, textiles, leather, paper and rubber and have an important role in safeguarding lives and property. Polymers such as thermoplastics and thermosetting resins, for example, epoxy resins and elastomers, have wide-ranging applications as floor coverings, fabrics, furniture, walls, ceiling fittings, paints, moulded products and polyurethane foams as well as in automotives, electronics and in electrical appliances and cables. Thermosetting resins are widely used in the manufacture of printed circuit boards and their connectors. Because most polymers are flammable, fire safety is important in these applications. It has been claimed that only 12% of plastics used today contain flame retardants.[1] In addition, about 10% of fires are caused by electrical faults in wiring and electrical equipment and that these fires account for 19% of fire-related injuries. Worldwide sales of flame retardants are approximately $4 billion per annum and this is expected the grow to $5.8 billion by 2018. When used in practice, the loadings of flame retardants should not affect the physical properties of the base polymers and the retardants should not produce toxic and corrosive combustion products. Flame retardants form a buoyant area of research as

Exploring Materials through Patent Information
By David Segal
© David Segal, 2015
Published by the Royal Society of Chemistry, www.rsc.org

there are concerns about the toxicity of low molecular weight brominated materials and there is requirement to find cheap and effective alternatives. An overview of flame retardants is given in this chapter, with particular reference to developments in the patent literature.

11.2 MODE OF OPERATION OF FLAME RETARDANTS

There are three mechanisms by which flame retardants work.[3] Brominated materials act within the gas phase of the flame, where low reactivity bromine free radicals scavenge highly energetic oxygen and hydroxyl free radicals derived from combustion products of the polymer, thus inhibiting flame propagation by free radicals. Phosphorus-based flame retardants act by cross-linking radicals produced when polymer chains are broken down forming a carbon-based crust or char that acts as a thermal barrier which shields the underlying polymer from the gaseous phase. Inorganic flame retardants such as aluminium hydroxide cool the reaction down by absorbing heat and releasing water vapour. The release of water vapour may cool combustible materials to below the temperature required to sustain combustion processes. Incorporation of inert fillers into flammable polymers can dilute the available fuel for combustion. Some retardants such as trichloroisopropyl phosphate combine the benefits of both halogen and phosphorus-based flame retardants.

11.3 MATERIALS FOR FLAME RETARDANTS

11.3.1 Nanoclays

Flame retardants are used in the microelectronics industry, for example, in mounting substrates including motherboards and expansion cards for use in personal computers.[4] While these have used bromine-based and phosphorus-based compounds, concerns over halides have encouraged the search for new flame retardants. A composition for a mounting board consisted of bismaleimide triazine polymer that was melt blended with clay platelets with a particle size less than 50 nm. Alumino-silicate clay minerals with a particle size in this range are referred to as nanoclays and the clay content in the polymer was up to 15 weight %. Examples of the clays include the natural minerals montmorillonite, saponite and hectorite. The nanoclay dispersion is cured to form the substrate core. In a further application of nanoclays, cobalt particles were deposited onto the surface of clay platelets from cobalt acetate solution, after which carbon nanotubes were deposited by chemical vapour deposition onto the clay platelets at about 1073 K.[5] The clay-carbon nanotube composites were dispersed in a polymer, epoxy resin and polyurethane and the nanotubes were claimed to aid in the separation of individual clay platelets, that is exfoliation, and their dispersion throughout the polymer.

Modified clays, both natural and synthetic have been considered as flame retardants. For example, montmorillonite and a synthetic lithium aluminium double layered hydroxide were modified by intercalation with the C_{60} fullerene and imidazole, respectively.[6] Modified clays were incorporated separately into a thermosetting resin, bisphenol-A novolac epoxy and hardened pieces of clay-polymer composite obtained on curing at 463 K. The intercalated materials have the potential to release free radicals on heating that scavenge free radicals generated from burning of the clay-polymer composite, lowering the heat released from the composite (Table 11.1) and maintaining its integrity.

When well dispersed in a plastic, nanoclays can improve physical properties such as mechanical strength and act as a flame retardant. Commercially available nanoclays are intercalated with quarternary ammonium compounds to aid their dispersability and such organically modified clays are referred to

Table 11.1 Effect of a 3 weight % loading of modified clays on the heat released from a cured epoxy thermosetting resin.[6]

Sample	Heat released $(MJ\ m^{-2})$
Bisphenol-A novolac epoxy	98
Bisphenol-A novolac epoxy plus montmorillonite modified by C_{60} fullerene	94
Bisphenol-A novolac epoxy plus lithium aluminium double layered hydroxide modified by imidazole.	10

as organo-clays.[7] However, on exposure to flames the decomposition products of quarternary ammonium compounds can support the spread of flames. As a potential alternative to quaternary ammonium compounds, nanoclays have been coated with diphosphate-based compounds with boiling points greater than 573 K, in particular with resorcinol diphosphate and bisphenol diphosphate, which are sprayed as hot liquids onto the clay particles, such as montmorillonite or bentonite. In another approach intercalation of montmorillonite was intercalated with a hydrolysed n-aminoethyl-3-aminopropyl trimethoxysilane to encourage chemical bonding between the silane and hydroxyl groups on the clay, after which the clay slurry was filtered, dried and milled to a fine powder.[8] Decomposition products of the silane include silicon carbide and are not flammable. Other compounds used for intercalation include ethyltriphenylphosphonium bromide and tetrabromobisphenol A bis-2,3 dibromopropyl ether.

In general, the loading of flame retardants in a polymer ranges between 10 weight % and 30 weight %. For organic retardants higher loadings cause deterioration in the mechanical properties of the polymer without a beneficial effect on flame retarding properties. However, for inorganic retardants higher loadings increase the brittleness of the polymer but may improve the flame retarding effect. Good dispersion of the flame retardant throughout the polymer is essential to maximise the properties of the retardant and this can be achieved by intercalation and exfoliation of nanoclays. As an example, poly(oxyalkylene)amine was mixed with hexachlorocyclotriphosphazene to replace chlorine groups in the phosphazene and the resulting complex intercalated into monomorillonite by ion-exchange.[9] The ion-exchanged nanoclay was dispersed into thermoplastic polyurethane or rubber at concentrations up to around 30 weight %. Thermogravimetric analysis showed (Figure 11.1) that for polyurethane the residual carbon contents of samples was higher in the presence of intercalated montmorillonite, thus indicating that the latter retarded flame propagation.

Intercalated nanoclays are composite materials and as clay platelets and the intercalating materials are nanoparticles these composite systems may be referred to as nanocomposites.

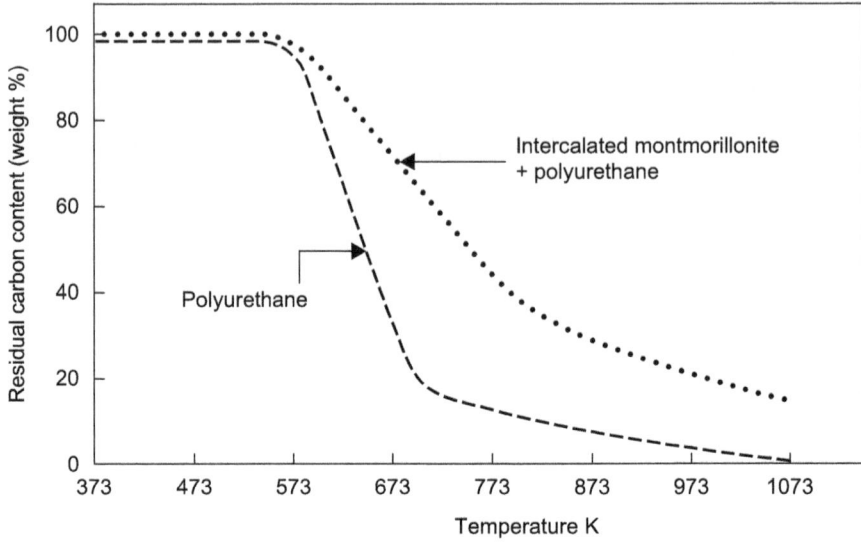

Figure 11.1 Effect of intercalated montmorillonite carbon content of polyurethane.[9]

11.3.2 Intumescent Coatings

The range of flame retardant materials cannot easily be separated into discrete categories, as was observed for nanoclays in which a retardant can combine an inorganic material and an organic compound. However, a class of flame retardants are intumescent coatings that form a protective barrier for substrates in the event of a fire. The coatings act as insulating thermal barriers, protecting the underlying substrate from heat generated in the fire. In use an intumescent flame retardant system expands to form a protective solid foam on the underling substrate. Applications for intumescent coatings include fire proofing internal walls of buildings such as timber panels and blocks.[10] Key components of a coating composition for an intumescent coating include a charring or carbonisation agent, pentaerythritol, a blowing agent, melamine and a flame retardant component such as ammonium polyphosphate; the coating composition is applied as a paint. For its application in intumescent paints, ammonium polyphosphate requires a low solubility because soluble material will be converted by moisture in the air to ammonium orthophosphate solution.[11] The generic

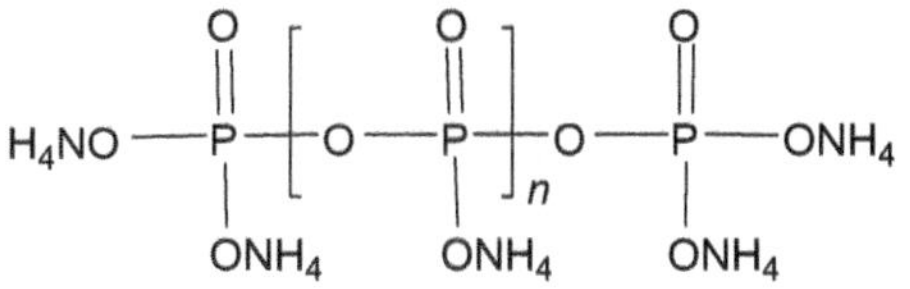

Figure 11.2 Generic chemical formula for ammonium polyphosphate.[11]

chemical formula for an ammonium polyphosphate $(NH_4PO_3)_n$ (n >10) is shown in Figure 11.2.

An industrial process for manufacture of ammonium polyphosphate with the crystal from II type suitable for use as a flame retardant involves mixing phosphorus pentoxide, diammonium orthophosphate $(NH_4)_2HPO_4$ and ammonium sulphate at 473 K for about 5 minutes, after which a polymerisation reaction takes place in the presence of ammonia gas at 980 Pa. The mixture is then homogenised and then a further polymerisation reaction takes place under ammonia using a carefully controlled temperature gradient that resulted in ammonium polyphosphate with low solubility in water, 0.002 g cm^{-3}.

11.3.3 Aluminium Hydroxide

The alumina trihydrate gibbsite, α-Al$_2$O$_3 \cdot 3H_2O$ is widely used as a fire retardant filler in glass-reinforced plastic composite materials.[12] Gibbsite can absorb heat and release water vapour from its decomposition on heating to alumina, thus reducing the rate of combustion of a polymer.[13] It is important to increase the loading of gibbsite without compromising the physical properties of the composite material and small particle sizes are required to achieve this objective. Gibbsite is obtained by precipitation from sodium aluminate in the Bayer process for applications including aluminium smelting but for this majority application fine powder sizes are not required. Fine gibbsite powders with diameters between 1.2 μm and 2.2 μm have been prepared by seeding Bayer liquors with bayerite crystals.[12] Expandable graphite is a form of graphite in which the lattice has been intercalated with an acid such as sulphuric acid, so that on heating the intercalated ions produce gaseous decomposition products that expand the lattice.[13] When added to a composite of a polymer, a thermoplastic or thermosetting resin with loadings up to 2 weight % of the gibbsite content, the expandable graphite produces a reduced thermal conductivity of the composite material resulting in a reduced rate of heat release.

11.3.4 Melamine Derivatives

Condensation products between melamine and a pentaerythritol ester of phosphoro chloridic acid have very low solubility in water compared to phosphorus-based flame retardants and good flame resistance when dispersed in a polymer such as polypropylene or applied to wood structures.[14] In a representative synthesis, pentaerythritol was heated with phosphorus oxychloride at up to 463 K to produce a pentaerythritol ester of phosphoro chloridic acid as shown schematically in Figure 11.3.

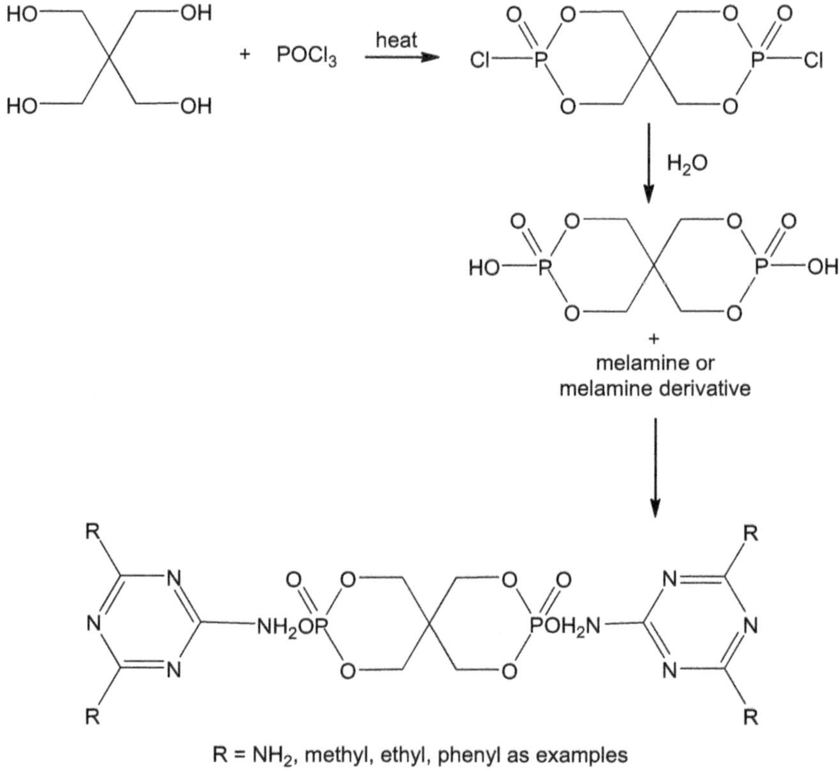

R = NH_2, methyl, ethyl, phenyl as examples

Figure 11.3 Melamine condensation product for use as a flame retardant.[14]

The pentaerythritol ester of phosphoro chloridic acid was dissolved in water that resulted in substitution of the Cl group by the OH group after which the substituted ester was reacted with melamine as indicated in Figure 11.3 yielding a dicyclic phosphorus-melamine compound. In a different approach to melamine-based fire retardants, a salt of a melamine condensation product and a phosphinic acid was obtained by reaction of melamine and ethylmethylphosphinic acid, $(CH_3)(CH_3CH_2)P(=O)OH$, at 563 K.[15] Ammonia was liberated and removed by a flowing nitrogen stream and after cooling the product contained 67 weight % melam and 1 weight % melamine.

11.3.5 Brominated Flame Retardants

Brominated compounds are incorporated into organic polymers and textiles for use as fire retardants. The polymers undergo a melt-processing operation during the manufacture of products and the flame retardant requires sufficient thermal stability to resist thermal degradation when present in a polymer melt at temperatures of 523 K or higher. The chemical formulas of two common brominated flame retardants, hexabromocyclododecane and pentabromodiphenylether are shown in Figure 11.4.[2]

There are concerns over the toxicity of low molecular weight brominated retardants and their potential ability to accumulate in body tissue. Brominated compounds and polymers with a molecular weight greater than 1500 are considered safer, with less tendency to bioaccumulate.[16] There is increasing regulatory and public pressure to find replacements for low molecular weight compounds such as hexabromocyclododecane.

The bromination of aliphatic unsaturated chemical groups has been used for the synthesis of brominated derivatives, but the reaction is exothermic and heat has to be removed effectively, especially for large-scale production, for both safety reasons and to avoid the production of by-products.[17] Removal of heat was achieved by using an inert solvent such as methylene chloride, whose evaporation during the reaction after adsorption of its latent heat of vaporisation terminated the reaction and prevented formation of by-products. Monomeric flame

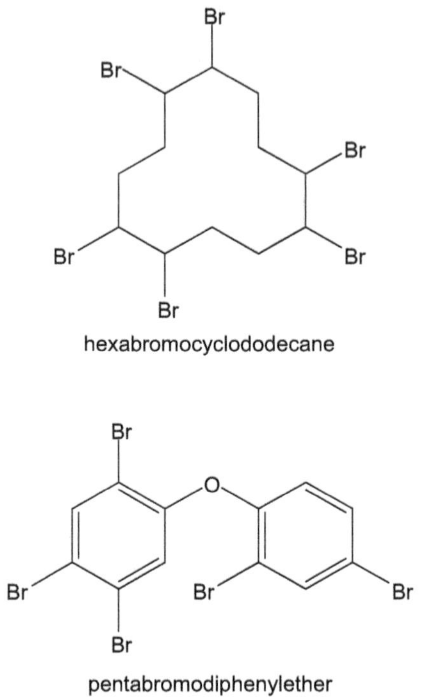

hexabromocyclododecane

pentabromodiphenylether

Figure 11.4   Examples of brominated flame retardants.[2]

retardants have been prepared by bromination of toluene in the presence of transition metal halide catalysts but bromination of benzyl bromide using a slurry of AlCl$_3$ catalyst in bromine at 333 K yielded a precipitated polymeric brominated poly-phenylmethane.[18] Bromine acted as a solvent but changing the solvent type, amount of catalyst and the amount of bromine affected the polymer characteristics, for example, its molecular weight.

Tetrabromobisphenol-A is used extensively as a flame retard-ant, particularly for styrene-based thermoplastics and processes are required for its production in high yield, with few by-products and with a light colour such as pale yellow.[19] Lightly coloured flame retardants are desirable as they have an aesthetic appeal when incorporated into products. An appropriate process involved reaction of bisphenol-A, hydrogen peroxide, H$_2$O$_2$ and bromine in ethyl alcohol at 333 K. The bromine feed was adjusted to maintain a stoichiometric ratio with bisphenol-A and the product, tetrabromobisphenol-A formed by precipitation. The presence of H_2O_2 in solution resulted in oxidation of HBr in solution back to bromine. The amount of unreacted bromine affected the colour of the precipitate and rapid formation of tetrabromobisphenol-A prevented precipitation of the intermediate tribromobisphenol-A. Polyhalogenated diarylalkanes, for example, decabromodiphenylethane are used as flame retardants for polyolefin and polystyrene-based formulations where only bromination of the aryl rings has been carried out.[20,21] A poly(bromoaryl)bromoalkane was obtained by dispersing a decabromodiphenylethane wet cake to a slurry in chlorobenzene and bromine, after which the mixture was refluxed at 406 K and then exposed to a light source. Photobromination yielded dodecabromodiphenylethane for use as a flame retardant.

Polybrominated aromatic compounds such as decabromodiphenyl ether and decabromodiphenyl ethane are produced (22) by brominating diphenyl ether and diphenyl ethane, respectively, in the presence of a Lewis acid, aluminium chloride in a solvent. The desired product, for example, decabromodiphenyl ether precipitates from solution but may contain significant quantities of underbrominated species, such as nonabromodiphenyl ether at a loading up to 2.5 weight %. However, underbrominated material cannot be easily separated from the desired product by re-crystallisation, as both the latter and nonabromodiphenyl ether have low solubility. When bromination of diphenyl ether was carried out in a solvent mixture of dichloromethane, bromochloromethane and dibromomethane at the same time as the precipitate was milled with ceramic beads. Decabromodiphenyl ether comprised 99.4 weight % of the product while the nonabromodiphenyl ether content was 0.1 weight %. The comminuted polybrominated compounds had a particle size of 7.1 μm. Polybromoaryl ethers have been prepared by the alkaline polymerisation of a brominated phenol, for example, tribromophenol in the presence of an initiator such as potassium persulphate.[23] A new class of fire retardants is based on continuous or batch bromination of linear styrenic polymers formed by anionic polymerisation of styrene monomer in the presence of a catalyst aluminium chloride using a short

bromination time of 20 minutes.[24,25] Short reaction times produced brominated compounds with desirable melt flow and
colour properties, while different reaction times resulted in different chemical structures for the brominated products.

11.3.6 Phosphorus-based Flame Retardants

The melting point for salts of phosphinic acid are listed in
Table 11.2 and the melting point for zinc methylethylphosphinic
acid is sufficiently low at 478 K, enabling incorporation of this
salt into a melt-processed polymer for use as a flame retardant.[26]

The chemical structure for zinc methylethylphosphinate is
shown in Figure 11.5.

Zinc dimethylphosphinate and zinc diethylphosphinate could
also be melt-processed with organic polymers such as blends
of acrylonitrile-butadiene-styrene, polyamides, polyesters and

Table 11.2   Melting point for salts of phosphinic acid.[26]

Metal	Phosphinic acid	Melting point (K)
Zn	Diethylphosphinic acid	482
Zn	Methylethylphosphinic acid	478
Ca	Methylethylphosphinic acid	>623 (decomposition)
Ba	Methylethylphosphinic acid	>623 (decomposition)
Mg	Methylethylphosphinic acid	>623 (decomposition)
Al	Methylethylphosphinic acid	>623 (decomposition)
Al	Diethylphosphinic acid	>623 (decomposition)

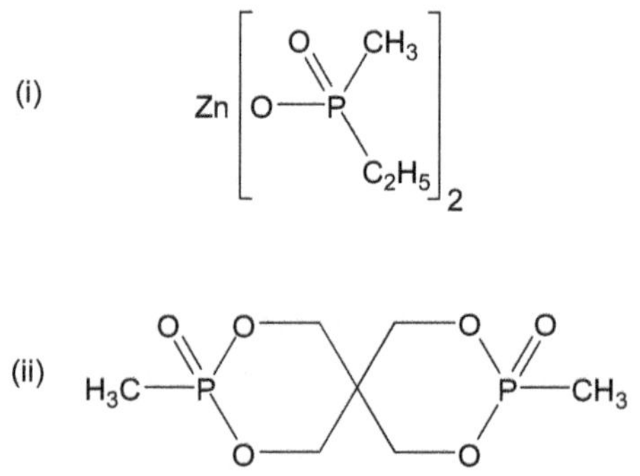

Figure 11.5   (i) Structure of zinc methylethyl phosphinate[26] and (ii) a cyclic
phosphonate.[27]

polycarbonates. Phosphorus-based flame retardants work by forming a protective char layer that acts as a thermal barrier and protects the underlying substrate from attack by oxygen and heat. A flame retardant composition consisting of a blend of two or more compounds that act together in a synergistic way has been developed.[27] For example, a mixture of a phosphonate compound (Figure 11.5) and a 1,3,5-triazine compound allowed lower concentrations in a polymer compared to the concentrations required for individual components without compromising the efficiency of the retardants. Flame retardants based on phosphonates can be hygroscopic and have a tendency to migrate from thermoplastic matrices over time.[28] In order to overcome these drawbacks, a combination of three halogen-free linear, branched or cyclic phosphonate compounds was formulated into a paint or varnish. Textiles could be treated with the combination of phosphonates. Flame retardants are used extensively in polyurethane foams and there is evidence that during the foam-forming reaction the hydroxyl group of organophosphorus flame retardants reacts with isocyanate groups in the foam-forming composition, thereby anchoring the flame retardant to the foam matrix.[29] Monohydroxy cyclic phosphonates such as 1,3,2-dioxaphosphorinane-5-methanol, 5-ethyl-2-(2-methyl)-, 2-oxide were suitable compounds for use as flame retardants for flexible polyurethane foam.

The flow properties of a polymer melt containing a flame retardant are important for shaping products and an objective in the development of novel flame retardants is to improve the properties of polymer melts without compromising flame retarding properties. Phosphate-based esters have been prepared from flavans or spirodichromans by reaction with phosphorus oxychloride, monophenyl dichlorophosphate or diphenyl chlorophosphate and could be incorporated as flame retardants into organic polymers such as polymethacrylates.[30] Flavans and spirodichromans could be synthesised from resorcinol. Hence, resorcinol reacted with acetone in the presence of hydrochloric acid catalyst to produce 2,4,4-trimethyl-2′,4′,7-trihydroxy flavan. Phosphinic acid esters have applications as chemical intermediates for flame retardants and have been synthesized and isolated in pure form.[31] Yellow phosphorus was reacted with an alkyl halide, methyl chloride under alkaline conditions to form methanephosphonous acid, $CH_3P(=O)H(OH)$. The latter was esterified and the ester reacted further with an olefin. Aryldiphosphate esters have applications as flame retardants and they have been prepared from intermediates formed by reaction of phosphorus oxychloride with a dihydric aromatic compound such as resorcinol, while non-halogen cyclic phosphate esters with linear or branched alkyl groups have been developed for use as flame retardants for polyurethane foams.[32,33]

Pressure sensitive adhesives are widely used and the potential flammability of acrylic adhesives has resulted in the incorporation of halogenated flame retardants.[34] These adhesives are polymerisation products between a monomer mixture, for example, a (meth)acrylate ester and vinyl carboxylic acid. However, the desire to phase out halogenated compounds has widened the search for alternative flame retardants. Cyclic phosphonate esters have been developed as flame retardants for acrylic pressure-sensitive adhesives with a loading of the retardant up to 15 weight % without reducing the adhesive performance. It is desirable for a flame retardant to melt during incorporation into a polymer to achieve good dispersion. Aluminium methyl methylphosphonate is a flame retardant that does not melt during processing with organic polymers but good dispersion can be obtained by incorporation into polymers as fine powders to give translucent or transparent polymer composites.[35] Aluminium methyl phosphonate was obtained with a particle size less than 10 μm by reaction of aluminium hydroxide with dimethyl methylphosphonate in the presence of a phase transfer catalyst, tetra-n-butyl phosphonium bromide at 483 K and had an acicular morphology. Polyester is widely used for the production of melt-spun textile fibres that incorporate flame retardants.[36] Polyphosphonate flame retardants with molecular weights in the range 10,000 to 120,000 have been developed for melt spinning with polyesters and the polyphosphonates had a lower tendency to leach out of polyester-based products than lower molecular species. Phosphorus chemistry allows a wide range of structures to be prepared for potential use as flame retardants. For example, neopentyl glycol hydrogen phosphonate was reacted with p-benzoquinone and acetic acid at 383 K to produce a novel flame retardant.[37] Aromatic diphosphates have been developed as flame retardants for thermoplastic and thermosetting resins.[38] For example, an aromatic monohydroxy compound such as 2,6-xylenol was heated in the presence of a magnesium chloride catalyst at 393 K after which phosphorus oxychloride was added to the mixture and the temperature increased to 353 K. Di-(2,6-xylyl)phosphoro chloridate was obtained and reacted with an aromatic dihydroxy compound such as resorcinol. A dehydrochlorination reaction took place at 453 K, yielding an aromatic diphosphate. Effective flame retardants for polyurethane foams have been developed by reaction between a polyol, a polyamine and a phosphorus compound such as 2-chloro-5,5-dimethyl-1,3,2-dioxaphosphorinane.[39]

Red phosphorus has been incorporated into polymer blends of polycarbonate and acrylonitrile-butadiene-styrene resins at loadings up to 10 weight %.[40] However, when used as a replacement for red phosphorus, triphenyl phosphate had little beneficial effect as a flame retardant as it acts as a liquid plasticiser because it has a low melting point, around 323 K, causing segregation of components. Epoxy resin sealing materials are used to protect integrated circuits from moisture and dust.[41] Halogenated epoxy resins, alone or in combination with antimony trioxide, have been used as conventional flame retardants, but there are environmental pressures to replace halogenated retardants. Red phosphorus has been considered as a flame retardant for semiconductor sealing materials but, on exposure to moisture, elevated temperature, oxygen or alkali its decomposition products include phosphine and acids of mono- to pentavalent phosphorus. The acids can corrode circuitry on the integrated circuits. However, when red phosphorus particles, about 20 μm in size, were coated with an anhydrous zinc compound such as zinc oxide and the coated particles were then sealed by coating with a thermosetting resin, leaching of PO_4^- ions was reduced and the stabilised phosphorus particles could be incorporated into an epoxy resin as a flame retardant. Incorporation of phosphorous-based compounds as flame retardants into polyamides tends to produce reddish brown-coloured plastics. However, incorporation of red phosphorus with a diameter up to 200 μm and melamine polyphosphate into polyamide moulding compositions yielded pale-coloured plastics with flame retardancy for applications including plugs and plug connectors and mouldings for automotive vehicles, office equipment, sports equipment, consumer goods and medical equipment.[42] Efficient blending of a flame retardant and thermoplastic is important to obtain a uniform distribution of the retardant. An organic phosphorus oligomer, for example, resorcinol bis(diphenyl)phosphate, was blended continuously with polycarbonate in a screw extruder and this mixing method limited the degradation of mechanical properties of the mixture and the colour of the resin.[43] Economic production methods are required for efficient flame retardants. A continuous process for synthesis of 9, 10-dihydro-9-oxa-10-phosphaphenanthrene-10-oxide as a flame retardant for polyesters, polyamides and epoxy resins has been developed.[44] The method involved esterification of o-phenylphenol with phosphorus trichloride in the presence of a catalyst, zinc chloride at 343 K, after which the temperature was raised to 423 K to effect a cyclisation-hydrolysis reaction.

Hydrogarnets have the general formula $M^{II}_3M^{III}_2(OH)_{12}$ where M^{II} and M^{III} are divalent and trivalent cations, respectively. An example of a hydrogarnet is $Ca_3Al_2(OH)_{11.4}(SiO_4)_{0.15}$. Preparation of a synthetic hydrogarnet containing octahedral crystals with a size around 5 μm were obtained from the reaction of alumina trihydrate, calcium hydroxide and phosphoric acid under alkaline conditions at 368 K.[45] The product had the chemical formula $Ca_3Al_2O_{0.3}(OH)_{10.5}(PO_4)_{0.3}$ and this synthetic hydrogarnet had a higher initial decomposition temperature of 523 K compared to the commonly used inorganic flame retardant alumina trihydrate, which is around 473 K as measured by thermogravimetric analysis. The garnet has the potential to act as a flame retardant at higher temperature compared to existing inorganic materials and to be blended with polymers that have processing temperatures greater than 473 K.

Printed circuit boards that use thermosetting plastics or thermoplastics employ fillers such as inert silica particles as rheology modifiers as well as flame retardants. Conventional phosphorus-based flame retardants are based on phosphonates, phosphate esters and phosphinates and the general molecular structure of these materials is shown in Figure 11.6.

Phosphorus-modified silica particles have been prepared that act both as rheology modifiers and flame retardants.[46] For example, hydrosilated terminated inorganic particles such as silica

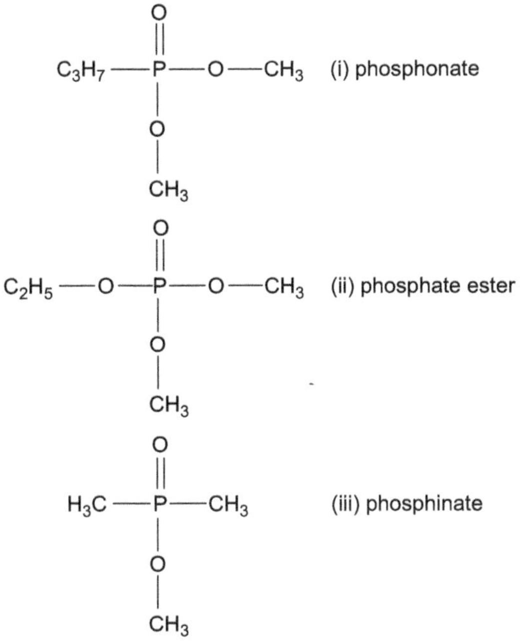

Figure 11.6 Molecular structures for (i) phosphonates, (ii) phosphate esters and (iii) phosphinates.[46]

particles are produced by first reacting the inorganic material with a coupling agent, triethoxysilane at room temperature. Then phosphorus-modified silica was obtained by coupling a vinyl-terminated phosphorus-based monomer, for example, dimethyl allylphosphonate onto the hydrosilated terminated silica. The printed circuit board contains a stack of conductive planes separated by a dielectric material. The phosphorus-modified silica flame retardant is incorporated into a varnish coating based on an epoxy resin and applied as a coating to the dielectric layers. Other phosphorus-based materials for use as flame retardants in epoxy resins have been prepared from benzoxazine ring-containing materials.[47] For example, from the reaction between 9, 10-dihydro-9-oxa-10-phosphaphenanthrene-10-oxide and a phenol-formaldehyde condensation product at about 473 K. Precursors were commercially available, which is an important factor when producing flame retardants on a large scale.

11.3.7 Ionic Liquids as Flame Retardants

The wide range of applications for ionic liquids was highlighted in Chapter 10 and a further potential application for them is as flame retardants, as they are considered to show resistance to migration and leaching within products, have low toxicity and do not accumulate in fatty tissue.[48] In addition, they have the ability to be prepared in a biodegradable form and hydroxymethyl imidazolium ionic liquid derivatives obtained from fructose have potential for fighting forest fires. 1-butyl-3-methylimidazolium bromide was melt blended with polyoxymethylene, while tri-ethylmethylphosphonium dibutyl phosphate was used in an intumescent flame-retarding system for polypropylene.

11.4 SUMMARY

Worldwide sales of flame retardants are approximately \$4 billion per annum, with an estimated growth to \$5.8 billion in 2018. Flame retardants form a buoyant area of research as there are legislative concerns over the toxicity of brominated materials and requirements to identify alternative materials. Materials used for flame retardants include nanoclays, brominated compounds, alumina trihydrate, melamine-based derivatives, phosphorus-based compounds and ionic liquids.

REFERENCES

1. J. L. Lee, Halogen-free flame retardant material, *United States Patent Application*, 2011/0144244, 2011.
2. N. Moran, Phasing out fire retardants, *Chem. World*, 2013, **10**(8), 50–53.
3. M. B. Block, J. Ferbitz, C. Fleckenstein, A. Cristadoro, B. Bruchmann and K. Massonne, Polymeric flame retardant, *United States Patent Application*, 2012/0010312, 2012.
4. P. Bhimaraj and O. Bchir, Nanoclays in polymer compositions, articles containing same, processes of making same and systems containing same, *United States Patent*, 8 163 830, 2012.
5. V. P. Veedu and V. Kamavaram, Hybrid nanocomposite for fire retarding applications, *United States Patent*, 8 173 734, 2012.

6. T.-Y. Tsai, S.-T. Lu, C.-H. Liu, C.-C. Huang, H.-C. Tsai and J.-C. Lin, A modified clay for use in a fire retardant polymer composite, *United Kingdom Patent Application*, 2 461 172 A, 2009.

7. D. Abecassis, Novel flame retardant nanoclay, *United States Patent Application*, 2008/0023679, 2008.

8. S. Kenig, Nanoclays containing flame retardant chemicals for fire retardant applications, *International Patent Application*, 2013/034954, 2013.

9. J.-J. Lin and T.-K. Huang, Phosphorus flame retardant containing clay, *United States Patent Application*, 2013/0165562, 2013.

10. D. A. Ward, Fire retardant coating, *United Kingdom Patent Application*, 2 433 938 A, 2007.

11. V. M. Fibla and L. T. Dalmau, Process for manufacturing crystal form II solid ammonium polyphosphate, *European Patent*, 0 555 168 B, 1999.

12. G. Bilandzic, N. Brown and N. Putz, Process for the production of aluminium hydroxide, *United States Patent*, 6 887 454, 2005.

13. G. C. Rex and M. A. Herndon, Improved use of alumina trihydrate with expandable graphite in composites, *International Patent Application*, 2014/018521, 2014.

14. D. H. Lee, D. H. Hyun, S. H. Gwon, H. D. Cho and S. B. Kim, Method for manufacturing dicyclic phosphorus melamine compounds having superior fire retardancy and fire retardant material using thereof, *United States Patent Application*, 2004/0236132, 2004.

15. W. Heinen, Salt of a melamine condensation product and a phosphorus-containing acid, *European Patent*, 1 252 168 B, 2005.

16. E. B. Vogel, S. L. Kram, D. J. Murray and B. A. King, Brominated and epoxidized flame retardants, *United States Patent*, 8 569 424, 2013.

17. Y. Taketani, H. Hoshimi, M. Monri, S. Tanabe and Y. Shimidzu, Bromine compound production method, *United States Patent*, 5 998 674, 1999.

18. S. Hussain, Brominated polyphenylmethanes, process for their preparation and fire retardant compositions containing them, *United States Patent*, 6 063 852, 2000.

19. T. Manimaran, H. Y. Einagar, R. A. Holub, A. E. Harkins, Jr. and B. G. McKinnie, Process for the preparation of tetrabromobisphenol-A, *United States Patent*, 6 300 527, 2001.
20. R. B. Dawson and S. Hussain, Poly(bromoaryl)alkane additives and methods for their preparation and use, *United States Patent*, 6 743 825, 2004.
21. R. B. Dawson and S. Hussain, Poly(bromoaryl)alkane additives and methods for their preparation and use, *United States Patent*, 7 129 385, 2006.
22. H. Stollar, G. Miaskovski, A. Meirom, M. Peled, J. Yu and I. B. David, Process for preparing polybrominated compounds, *United States Patent Application*, 2009/0227816, 2009.
23. L. D. Timberlake, W. R. Fielding, S. Mathur and M. V. Hanson, Brominated flame retardant, *United States Patent*, 7 718 756, 2010.
24. C. H. Kolich, J. T. Aplin and J. F. Balhoff, Brominated styrenic polymers and their preparation, *United States Patent*, 8 168 723, 2012.
25. W. J. Layman, Jr., C. H. Kolich, A. G. Mack, J. P. McCarney, G. Kumar, J. Morice, Z. Ge, B. Liu, D. W. Luther, K. M. White, J. Wang, R. R. Joshi and H. B. Chew, Branched and star-branched styrene polymers, telomers and adducts, their synthesis, their bromination and their uses, *United States Patent*, 8 476 373, 2013.
26. M. Sicken, E. Schlosser, W. Wanzke and D. Burghardt, Fusible zinc phosphinates, *United States Patent Application*, 2004/0176506, 2004.
27. V. Butz, Flame-retardant composition comprising a phosphonic acid derivative, *United States Patent Application*, 2014/0005289, 2014.
28. V. Schanen, D. Shamblee, G. Prlimov and J. A. Salter, Mixed phosphonate flame-retardants, *United States Patent*, 8 278 375, 2012.
29. A. Piotrowski, A. N. Desikan and S. Dashevsky, Monohydroxy cyclic phosphonate substantially free of polyhydroxy phosphonate, process for making same and flame retardant flexible polyurethane foam obtained therefrom, *United States Patent Application*, 2013/0237623, 2013.

30. R. B. Durairaj and G. A. Jesionowski, Phosphate ester flame retardants from resorcinol-ketone reaction products, *United States Patent*, 7 365 113, 2008.
31. S. Horold, N. Weferling, H.-P. Breuer and M. Sicken, Process for preparing phosphinate esters, *United States Patent*, 6 278 012, 2001.
32. J. S. Stults, Process to prepare aryldiphosphoric esters, *International Patent Application*, 97/47631, 1997.
33. J. Stowell and W. Liu, Phosphate ester flame retardant and resins containing same, *United States Patent Application*, 2010/0137465, 2010.
34. J. Buettner, E. Pyun, A. R. Piepys, L. S. Lim, F. C. D'Haese and T. Pick, Pressure sensitive adhesives containing a cyclic phosphonate ester flame retardant, *United States Patent Application*, 2013/0236718, 2013.
35. J. Zilberman, S. V. Levchik, A. Gregor, P. Georiette, Y. B. Yaakov and G. Shikolski, Metal phosphonate flame retardant and method producing thereof, *United States Patent Application*, 2013/0165564, 2013.
36. M. A. Lebel, L. Kagumba and P. Go, Phosphonate polymers, copolymers and their respective oligomers as flame retardants for polyester fibers, *International Patent Application*, 2012/068264, 2012.
37. A. Worku, A. Bharadwaj, R. Thibault, M. Wilson and D. Potts, Phosphorus-containing compounds and polymeric compositions comprising same, *International Patent Application*, 2010/025165, 2010.
38. K. Hirao, K. Ohtsuki, H. Tsuji and H. Sato, Method for producing aromatic diphosphates, *European Patent Application*, 2 592 087 A, 2013.
39. Y. Qi and X. Tal, Phosphorus-containing flame retardants for polyurethane foams, *International Patent Application*, 2012/126380, 2012.
40. H.-C. Kao, W.-J. Chen and W.-F. Kuo, Nonhalogen flame-retardant polycarbonate compositions, *United States Patent*, 5 436 286, 1995.
41. Y. Kinose, R. Imamura, A. Inoue, T. Hata and E. Okuno, Red-phosphorus base flame retardant for epoxy resins, red phosphorus-base flame retardant compositions therefor, processes for the production of both, epoxy resin

compositions for sealing for semiconductor devices, sealants and semiconductor devices, *United States Patent*, 6 858 300, 2005.

42. A. Konig, T. Erdmann, M. Roth, K. Uske, J. Engelmann, A. Ebenau and M. Klatt, Pale-colored flame-retardant poly-amides, *United States Patent Application*, 2013/0248782, 2013.

43. A. Miyamoto, K. Shibuya, H. Hachiya, C. N. Wu and W.-Y. Su, Method for producing a flame retardant polycarbonate resin composition, *United States Patent*, 6 833 397, 2004.

44. A. Kaplan, R. Gisler and C. Kohl, Process for continuously preparing an organophosphorus compound and use thereof, *United States Patent Application*, 2010/0298470, 2010.

45. M. Giesselbach, W. Hoepfl, R. G. E. Herbiet and G. P. Heines, Synthetic inorganic flame retardants, methods for their preparation and their use as flame retardants, *United States Patent Application*, 2011/0213065, 2011.

46. D. J. Boday, J. Kuczynski and R. E. Meyer, III, Non-halogenated flame retardants, *United States Patent Application*, 2013/0206463, 2013.

47. J. Gan, Phosphorus-containing compounds useful for mak-ing halogen-free, ignition-resistant polymers, *United States Patent Application*, 2014/0051796, 2014.

48. Y. Xu, Ionic liquid flame retardants, *United States Patent Application*, 2011/0073331, 2011.

CHAPTER 12

Graphene

12.1 INTRODUCTION

Studies on materials do not routinely attract attention in the public domain, but occasionally a material or process captures the public attention. For example, the discovery of high-temperature ceramic superconductors in 1986 resulted in unprecedented worldwide research interest throughout 1987.[1] Thousands of participants attended international conferences, the ceramics were frequently described in national newspapers alongside a growth in scientific articles and national initiatives were set up in several countries. The almost frenzied interest arose from potential applications of superconducting ceramics and the temperature at which they develop superconductivity. Anyone who fabricated a small pellet of the ceramic super-conductor and observed its levitation above a magnet immersed in liquid nitrogen due to the expulsion of magnetic fields in the Meissner effect could easily have been caught up in the excitement of the discovery of this new class of material. The interest and excitement in these materials arose from their potential applications due to the absence of electrical resistance including magnetically levitated trains, power transmission without energy losses and compact, powerful and efficient electric motors. Nowadays, the technique of three-dimensional printing

Exploring Materials through Patent Information
By David Segal
© David Segal, 2015
Published by the Royal Society of Chemistry, www.rsc.org

(Chapter 6) in which components are built up layer-by-layer is attracting much attention in the public domain and surgical operations involving implants made by this process have been widely reported, although 3D printing is not a new technique. Graphene, a two-dimensional sheet of carbon that is one atomic layer thick, is the thinnest known material and is attracting much publicity in the public domain. It has been described as a 'wonder material' and was isolated in 2004 at the University of Manchester by Sir Andre Geim and Sir Konstantin Novoselov.[2] The excitement surrounding graphene arises from its potential applications based on its electronic and mechanical properties, namely, electrical conductivity, electron mobility, mechanical strength, stiffness, thermal conductivity, flexibility and optical transparency. These applications include components of electrodes for fuel cells, lithium-ion batteries and supercapacitors, conductive transparent coatings for the replacement of indium tin oxide, integrated circuits, field effect transistors, optoelectronics, solar cells, a replacement for silicon-based semiconductor devices, flexible display devices, protective coatings and fillers or reinforcements for composites to modify mechanical, electrical and thermal properties. An overview of the development of graphene as viewed from the patent literature is described below.

12.2 CARBON STRUCTURES

Carbon exists in several crystalline allotropic forms, namely diamond, graphite, fullerenes and carbon nanotubes and all of these allotropes have unique crystalline structures.[3] Fullerenes are also known as 'buckeyballs', with a surface pattern similar to the surface of a football. Carbon nanotubes are cylindrical structures grown with a single wall or multi-wall structures and have diameters ranging from a few nanometres to several hundred nanometres. If a carbon nanotube is cut along its length and its wall is flattened then a one-atom-thick planar sheet of carbon atoms that are densely packed in a crystalline honeycomb hexagonal lattice is obtained (Figure 12.1).[3]

This planar two-dimensional sheet represents a monolayer of graphene that is the basic structural element of graphite, fullerenes and carbon nanotubes. Graphene is not an allotrope

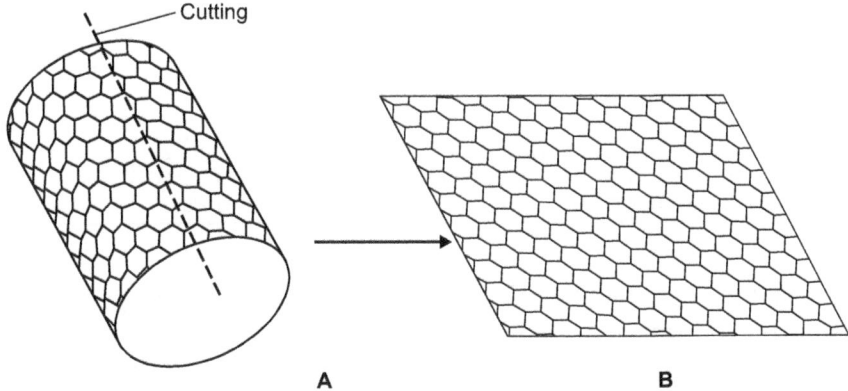

Figure 12.1 Conversion of a carbon nanotube (A) to monolayer graphene (B).[3]

of carbon because, as its thickness is one carbon atom layer thick, it does not form a three-dimensional crystal. Structurally, graphene has hybrid orbitals formed by sp^2 hybridisation. In sp^2 hybridisation the 2s orbital and two of the three 2p orbitals of a carbon atom mix to form three sp^2 orbitals and the one remaining p-orbital forms a pi-bond between the carbon atoms.[4] The structure of graphene is similar to the structure of benzene and the carbon-carbon bond length in graphene is approximately 0.142 nm.[5] A perfect graphene structure consists exclusively of hexagonal cells and any pentagonal or heptagonal cells constitute structural defects. Introduction of heptagonal or pentagonal cells into the graphene structure deform the planar structure into complex shapes. Graphite is a three-dimensional stacked structure, carbon nanotubes are one-dimensional structures, while fullerenes are zero-dimensional structures. It is the mechanical and electronic properties of graphene shown in Table 12.1 that have attracted considerable attention, as they offer the potential for a wide range of applications.[6] In particular, the higher electron mobility compared to conventional semiconductor materials offers the potential for higher computing speeds. However, the magnitude of the properties listed in Table 12.1 refer to an ideal defect-free graphene monolayer. While it is the goal of much experimental work to produce graphene monolayers, the unique properties of graphene are lost if experimental techniques result in a graphene nanoplatelet

Table 12.1 Properties of graphene.[6]

Property	Value
Young's modulus	1,100 G Pa
Thermal conductivity	5,000 W m^{-1} K^{-1}
Theoretical specific surface area	2,630 m^2 g^{-1}
Electron mobility at room temperature	200,000 cm^2 V^{-1} s^{-1}
Tensile strength	125 G Pa

consisting of a few layers of graphene so that the properties of multi-layer graphene tend towards those of bulk graphite.

The room temperature thermal conductivity, 5,020 W m^{-1} K^{-1}, and electrical conductivity, 6 S cm^{-1}, are approximately ten times higher than the respective values for silver.[7] The initial isolation of graphene in 2004 was achieved in a mechanical exfoliation in which an adhesive tape such as SCOTCH™ tape is applied against a graphite surface that has graphene layers.[8] The graphite layers attach to the SCOTCH™ tape that is removed from the graphite surface, after which the graphene layers are removed from the tape.

12.3 PREPARATION OF GRAPHENE

12.3.1 Wet Chemical Methods

One general approach to the synthesis of graphene involves carrying out oxidation-reduction reactions in a liquid medium. For example, in the Hummers method graphite powder is oxidised by a mixture of anhydrous sulphuric acid, an anhydrous nitrate salt and anhydrous potassium permanganate to produce graphite oxide.[9] The latter has surface epoxy groups as illustrated in Figure 12.2.[10]

Graphite oxide is not electrically conducting, as the surface modification destroys the sp^2 hybridisation and can be dispersed or exfoliated in a liquid medium to produce individual sheets of graphene oxide. Electrical conductivity can be restored by using a reducing agent such as hydrazine monohydrate to remove surface epoxy groups resulting in a dispersion of graphene. For clarification, graphene oxide may be considered to be an exfoliated or delaminated graphite oxide. In a modification of the oxidation method, graphite oxide was prepared by a

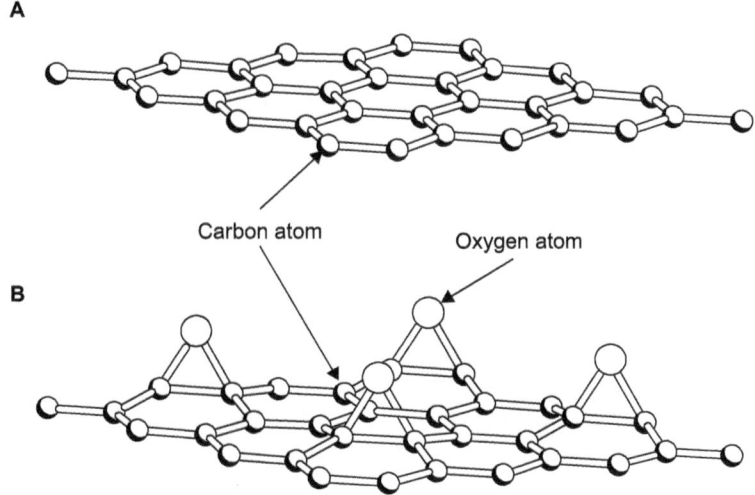

A

Carbon atom   Oxygen atom

B

Figure 12.2   Chemical structures for graphite (A) and graphite oxide (B).[10]

modified Hummers process using a mixture of sulphuric acid, sodium nitrate and potassium permanganate, after which it was ultrasonically dispersed in a mixture of trifluoroacetic acid, CF_3COOH, and hydriodic acid, HI, to a dispersion of graphene oxide.[11] Reduction of graphene oxide to graphene by HI took place at 263 K and because CF_3COOH is a weaker acid than HI, over reduction to a saturated hydrocarbon with a sp^3 hybridisation was avoided. In addition, whereas reduction with hydrazine introduces nitrogen into the reduced graphene oxide, with a subsequent reduction in electrical conductivity, the reducing mixture for this invention produced graphene with a high chemical purity. In another wet chemical method, graphite was first oxidised with a solution of potassium dichromate in nitric acid.[12] However, instead of reducing the graphite oxide with a reducing agent such as hydrazine, graphite oxide was washed, dried and then stripped of surface groups by calcination at temperatures between 773 K and 1273 K, after which the product from calcination was extensively ball-milled and ultrasonically dispersed in ethyl alcohol. Avoidance of a wet chemical reduction step produced graphene of high chemical purity with potential applications including lithium-ion batteries, supercapacitors, composite materials, transparent conductive films and microelectronic devices. Ball-milled material could be dispersed in non-aqueous solvents such as n-methyl pyrrolidone. A modified Hummers method avoids the use of viscous anhydrous sulphuric acid by adding limited amounts of water to the oxidising mixture.[13]

Exploitation of graphene dispersions for coating applications requires the dispersions to be stable, with no aggregation of graphene particles to larger clusters. Graphene dispersions fall within the area of colloid chemistry and can be described as colloidal dispersions. One way in which stable dispersions of graphene have been achieved is by dispersing graphite oxide into water in the presence of a dispersing agent such as sodium dodecylbenzene sulphonate.[14] The addition of ascorbic acid to the dispersion chemically reduced the graphite oxide to graphene particles that remained well dispersed. An alternative approach to obtaining well-dispersed graphene was to first exfoliate graphite oxide by heating the powder at 1373 K in a vacuum for 1 minute.[14] The resulting graphene powder was dispersed in water containing sodium dodecylbenzene sulphonate, after which ascorbic acid was added to reduce the surface oxygen content of the graphene particles. A grey graphite oxide powder was obtained by ultrasonically dispersing graphite powder in a mixture of sulphuric acid and nitric acid to oxidise the particle surface after which the powder was washed with ethyl alcohol and dried.[15] The dry powder was dispersed in N, N-dimethylformamide and then mixed with a solution of zinc acetate in N, N-dimethylformamide and heated at 368 K for 5 hours, a process that produced a zinc oxide-graphene core-shell structure with a particle size around 10 nm Powders with a copper oxide-graphene core-shell structure could also be produced.[15]

Reduced graphene oxide foams have been prepared for applications including supercapacitors and absorbents for liquids such as lubricating oils.[16] A graphene oxide dispersion was filtered through a porous membrane to from a compact that was exposed to the vapour of a reducing agent such as hydrazine monohydrate in an autoclave at 363 K for 10 hours. Gases released from hydrazine reduced graphene oxide to graphene and caused the compact to 'rise' and expand to form a porous solid. The tensile strength of the foams was around 3.2 M Pa.

The use for graphene in composite materials requires graphene in a form that can be easily handled. Hence a graphite source such as expanded graphite was reduced to the intercalation compound $K(THF)_yC_x$ by a solution of potassium naphthalate in tetrahydrofuran (THF).[17] The dried compound was dissolved in a polar aprotic solvent, N-methyl pyrrolidone and the solution consisted of graphene platelets. The graphene solution had potential uses for coating substrates such as silicon and in the fabrication of composites by mixing with a polymer. Another intercalated compound was obtained by dispersing nano-graphite platelet fibres in a reducing solution of potassium dissolved in liquid ammonia.[18] The dried and reduced graphite dispersion was dissolved in tetrahydrofuran to produce a non-agglomerated dispersion of graphene nanoplates.

Graphene fibres have been prepared.[19,20] For example, supporting fibres were first produced from nylon-6 or chitosan, after which a graphene oxide composite fibre was obtained by dip coating.[19] After the coated fibre was washed and dried, the graphene oxide coating was reduced to graphene at room temperature by vapour from a mixture of hydriodic acid and acetic acid. The supporting fibre was then removed by heating between 373 K and 473 K. In another approach,[20] graphite oxide was reduced with hydrazine and the resulting graphene flakes that had surface COOH groups were dispersed in water and the dispersion was then mixed with the surfactant sodium dodecylbenzene sulphonate to prevent particle aggregation. The stabilised dispersion was injected into a bath containing polyvinyl alcohol and composite fibres were obtained by wet spinning. The composite graphene/polyvinyl alcohol fibre was annealed st 873 K to remove polyvinyl alcohol, a process that produced porous graphene fibres.

12.3.2 Vapour Phase Methods

Chemical vapour deposition (CVD) has been used to fabricate graphene layers.[21] For example, a 300 nm thick layer of nickel was deposited by electron beam evaporation onto a silicon wafer containing a silica coating. A gaseous mixture of methane, hydrogen and argon was passed over the Ni surface that was kept at 1273 K. A graphene sheet with two to five layers of graphene deposited onto the nickel and was separated from the Ni by immersion in ferric chloride solution and then transferred to a silicon wafer. Intercalated graphene was produced by heating the graphene sheet to 523 K and then exposing the graphene to potassium vapour. Single layer graphene separated by potassium layers was identified from X-ray diffraction. Intercalated layers were also obtained with arsenic pentafluoride and ferric chloride. Graphene is a semimetallic material rather than a semiconductor, which limits its use in microelectronic applications.[22] A bandgap has to be introduced into graphene to facilitate its conversion to a semiconductor and use in integrated circuits. Graphene was grown on a Ni substrate by chemical vapour deposition, after which the graphene surface was subjected to a dry oxidation process using either an oxygen plasma or an ozone/ultraviolet treatment. Adsorbed oxygen species introduce defects into the sp^2 structure of graphene and disrupt the pi-bond network, affecting the electron mobility and charge transport across the graphene plane. A maximum bandgap in graphene of 2.5 eV corresponded to an oxygen concentration of around 21%.

Transition metals, nickel, iron, cobalt, copper, platinum, iridium, ruthenium and gold are useful for growing nanotubes and graphene, due to their ability to form carbides and the solubility of carbon in them at high temperature.[23] In the case of cobalt, the metal hydroxide $Co(OH)_2$ acts as the catalyst and was precipitated as flat platelets from cobalt chloride solution by ammonia released from hexamethylenetetramine solution on heating. The lateral dimension and thickness of the platelets was in the region of 1–2 μm and 40–50 nm, respectively. Cobalt hydroxide platelets were impregnated into a magnesium oxide support after which the supported catalyst was heated to 1073 K in argon in a furnace and exposed to an atmosphere of ethanol. The furnace was cooled to room temperature under an argon atmosphere and the magnesium oxide support and platelets were dissolved in hydrochloric acid. A key feature of this synthesis was that the deposited graphene occupied the whole planar area of the platelets, as indicated in Figure 12.3, while the morphology of the graphene was affected by the underlying morphology of the catalyst, hence nanoribbons when copper nanoribbons were used.

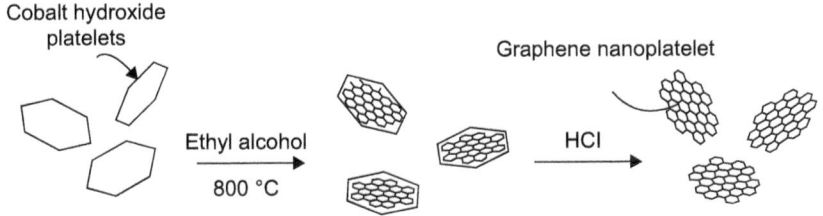

Figure 12.3   Formation of graphene on cobalt hydroxide platelets.[23]

Another approach for fabrication of graphene films is based on carbon segregation and the variation in the solubility of carbon in copper with changes in temperature, in particular the decrease in solubility from approximately 0.004 weight % to 0.002 weight % between 1223 K and 1073 K.[24] Thus, a copper substrate was exposed to a flowing stream of acetylene, C_2H_2, at 1223 K that resulted in absorption of C_2H_2 molecules onto the planar surface. Absorbed acetylene decomposed to carbon atoms that either remained on the planar surface or dissolved into the copper substrate, producing a homogeneous distribution due to the relatively high diffusion coefficient of carbon in copper, namely 3×10^{-11} m^2 s^{-1} at 1143 K. When the substrate was cooled to 1073 K, dissolved carbon atoms segregate at the planar surfaces of the copper substrate to form a homogeneous graphene film. Rapid cooling from 1073 K to ambient temperature inhibited further growth of the graphene film and the film could be removed by dissolving the copper substrate.

In another vapour phase deposition method, a silicon carbide layer with thickness in the range 10–100 nm was first deposited by sputtering on to a silicon wafer, after which a nickel film approximately three times the thickness of the silicon carbide layer was deposited on top of the non-oxide layer by sputtering (Figure 12.4).[25] The wafer was heated to a temperature in the range 623–1273 K that results in the formation of nickel silicide containing free carbon atoms. As the nickel silicide cools, carbon atoms migrate and precipitate at the surface forming a graphene layer. The nickel silicide layer could be dissolved in nitric acid solution leaving the graphene adhered to the support substrate.

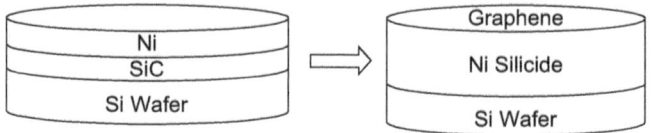

Figure 12.4 Graphene layer produced by carbon segregation.[25]

12.3.3 Use of Ionic Liquids

Wet chemical methods using oxidation-reduction reactions have been used to produce graphite oxide that is exfoliated to graphene oxide sheets by sonication in water and this approach aims to prevent aggregation of graphene sheets. An alternative approach to exfoliation has been developed by using ionic liquids.[26] For example, graphite powders less than 20 μm in size were mixed and ground with the ionic liquid 1-butyl-3-methylimadozolium hexafluorophosphate for up to 4 hours. The mixture was then added to a solution of N, N-dimethylformamide and acetone and centrifuged to remove the ionic liquid. The sediment was diluted in N, N-dimethylformamide and the supernatant dried down to a graphene sample. The graphene was in the form of nanosheets that were made up of two to four layers of graphene or nanodots. The latter had a diameter less than 10 nm and height between 1 nm and 3 nm. Graphene-like fragments of other layered materials could be obtained by methods similar to that used to prepare graphene, hence for MoS_2, WS_2, $MoSe_2$, $TaSe_2$, $MoTe_2$, $NiTe_2$, BN, Bi_2Se_3 and Bi_2Te_3. Exfoliation of graphite at concentrations between 0.01 weight % and 1 weight % in a mixture of one or more ionic liquids has been achieved. For example, a mixture of 0.015 weight % graphite in 1-hexa 3-methylimidazolium chloride and a mixture of 0.016 weight % graphite in 1-deca 3-methylimidazolium chloride were separately sonicated.[27] Exfoliated graphite and/or graphene remained suspended in the supernatant for more than 6 months. Composites were prepared by mixing a homogeneous solution of polystyrene in dimethylformamide with a 0.01 weight % solution of exfoliated graphite and/or graphene in 1-butyl 3-methylimidazolium chloride. Polystyrene which precipitated out on mixing had incorporated exfoliated graphite and/or graphene. Stable graphene dispersions were obtained using polymeric ionic liquids.[28] For example, graphite oxide made by oxidation of graphite with a mixture of potassium permanganate, sodium nitrate and sulphuric acid formed a yellow-coloured dispersion of graphene oxide in water. Poly(1-vinyl-3-ethylimidazolium)bromide, an ionic liquid polymer, was added to the aqueous dispersion. Graphene oxide particles were reduced on addition of hydrazine, resulting in a black dispersion containing unagglomerated graphene, which was stabilised by the polymeric ionic liquid and did not precipitate out.

12.3.4 Use of Electrochemical Methods

Graphene and graphite nanoplates have been prepared in an electrochemical cell in which the electrolyte contains alkylammonium ions.[29] As an example, a piece of highly ordered pyrolytic graphite was used as an anode with a platinum wire as a cathode in a cell that used N-methylpyrrolidone as a solvent containing tetrabutyl ammonium tetrafluoroborate $(C_4H_9)_4NBF_4$. Intercalation of cations into the graphite anode was confirmed by expansion of the electrode. Other suitable cations for intercalation included tributyl ammonium, $(C_4H_9)_3NH^+$ and diethyl ammonium, $(C_2H_5)_2NH_2^+$. Another electrochemical approach involved enclosing a carbon plate or powder in a polypropylene semipermeable membrane with a pore size of 40 nm and immersing this carbon electrode into sulphuric acid with a platinum wire electrode.[30] During electrolysis, exfoliation of the carbon electrode took place but only particles of graphene oxide with a size smaller than the pore size of polypropylene passed through the membrane, dried and reduced to graphene by heating at 723 K in an argon/hydrogen mixture for 30 minutes.

12.3.5 Doped Graphene

Interest in adding dopants to graphene to produce a functionalised material arises because of the potential to modify the properties of graphene. Fluorinated graphene has been referred to as fluorographene.[31] Graphene does not have a bandgap, but for applications in semiconductor devices it is necessary to engineer electronic properties such as a bandgap. Properties of graphene can be tailored by the introduction of dopants and the resulting functionalisation. Sources of fluorine include gaseous fluorinating agents, thus boron trifluoride, BF_3, nitrogen trifluoride, NF_3 and silicon tetrafluoride, SF_4 for examples, liquid fluorinating agents such as bromine trifluoride, BrF_3 and arsenic trifluoride, AsF_3 as well as solid sources of fluorine, hence zinc difluoride, ZnF_2 and chromium tetrafluoride, CrF_4. Fluorination is carried out at temperatures up to 723 K. Fully fluorinated graphene is a two-dimensional analogue for polytetrafluoroethylene and fluorination introduces a band gap into graphene. The ratio of fluorine to carbon was at least 0.50:1 corresponding to 50 atom % for the amount of fluorine bonded to graphene. Fluorographene has also been prepared by a solid-state reaction between graphite oxide and an inorganic compound such as ammonium fluoride or potassium fluoroaluminate or by heating polymeric materials, for example polyhexafluoropropylene with graphite oxide.[32]

Graphite oxide prepared by an aqueous oxidation-reduction method was dispersed in water and sonicated with cyanamide after which the mixture was dried and heated under argon at 823 K for 4 hours to convert the cyanamide to polymeric carbon nitride (C_3N_4).[33] The temperature was raised to a maximum of 1273 K to decompose the polymeric intermediate resulting in a nitrogen-doped graphene. The nitrogen content in the graphene was up to 12 weight % and other nitrogen-containing sources, for example, melamine, urea and dicyandiamide could be used. Metals including iron and cobalt could be introduced into the graphene lattice from salt solutions at the same time as graphene was mixed with cyanamide. Doped-graphenes have potential applications as electrode materials for supercapacitor and fuel cells particularly as replacements for platinum-based materials. Electronic devices (*e.g.* components for transistors, diodes, resistors, capacitors, inductors or sensors) have been prepared, in which one region of a graphene substrate is chemically functionalised with a dopant and another region is chemically functionalised with a different dopant, so that the two regions have distinct electronic properties.[34] As an example, an epitaxially grown graphene layer on nitrogen-doped silicon carbide has an intrinsic n-type character. If gold is deposited on a region of the graphene surface and bonded to the surface to form a p-type region then a p-n junction can be produced.[34]

When graphene is used as a filler for polymer composites, the presence of surface oxygen groups on exfoliated graphite oxide can have a deleterious effect on the chemical compatibility of graphene with the polymer matrix.[35] A carbon:oxygen molar ratio of at least 23 : 1 on functionalised graphene sheets was obtained by heating graphite oxide at s temperature of 1023 K or higher in a reducing atmosphere such as hydrogen diluted in argon. A further approach to producing functionalised graphene involved exfoliating graphite flakes by treatment with anhydrous sulphuric acid (oleum) and after repeated washing to remove residual acid, the graphite oxide was intercalated with tetrabutylammonium hydroxide, $(C_4H_9)_4NOH$, that resulted in a stable suspension.[36] Other graphitic structures including pentacene, rubrene and coronene that are almost insoluble in water and solvents. These materials were treated with oleum and intercalated with $(C_4H_9)_4NOH$. The functionalised materials were soluble in water and other solvents. There is a continuing demand for improved electrodes for batteries and doped-graphene has applications in this area. Evaporation of molybdenum oxide onto a graphene film grown by chemical vapour deposition and supported on a silica substrate followed by heating at 723 K lowered the resistance of the film allowing its potential use as a transparent electrode.[37]

12.4 LARGE-SCALE ROUTES TO GRAPHENE

Successful exploitation of graphene requires efficient production on a large scale and examples of potential manufacturing routes are summarised here. Graphite powder was mixed with N-methyl-2-pyrrolidone and subjected to supercritical carbon dioxide at 308 K that caused N-methyl-2-pyrrolidone to permeate into the graphite powder.[38] Rapid depressurisation of the CO_2 supercritical fluid exfoliated the graphite to graphene layers. Conversion of graphite to graphite oxide by oxidation-reduction reactions in an aqueous medium has been widely used but can involve the generation of wash liquors, an undesirable feature for large-scale production. Electrodialysis has been suggested as a way of treating liquids derived from the oxidation of graphite in solution in order to remove anions and cations from solution.[39] This method limits the amount of waste liquid generated during preparation of graphite oxide. Another method involves the use of vapour phase epitaxy to grow graphene films on a silicon carbide substrate at a temperature above 1373 K.[40] A key feature of this synthesis was the flow of argon over the substrate at a controlled flow rate in order to generate a stagnant boundary layer that prevented sublimation of silicon from the SiC substrate. Graphene layers with a controlled thickness were obtained by a CVD process in which propane molecules diffuse through the stagnant gas and decompose at the substrate surface. Uncontrolled growth of graphene films by migration of carbon to the surface of silicon carbide that occurs during sublimation of silicon is avoided.

12.5   TRANSFER METHODS FOR GRAPHENE

Chemical vapour deposition has the potential for large-scale manufacture of graphene. For example, a metallic catalyst layer, nickel, is deposited onto a silicon wafer containing a surface silica layer by, for example, e-beam evaporation.[41] Then the silicon wafer is exposed to a gas such as CH_4 that is subjected to an inductively coupled plasma. Carbon is absorbed into the nickel catalyst and the graphene film is grown and crystallised by rapid cooling of the catalyst. There are requirements to remove the graphene in an efficient way and this has been achieved by using steam to separate the oxide layer from the catalyst layer.[41] Graphene is grown on a thermal release tape that is wound around a roll in a roll-to-roll coating apparatus and the catalyst film is removed by etching. In another approach, the graphene film is peeled off from the nickel layer, an adhesive film is attached by a roller onto the graphene surface and the graphene is transferred to the adhesive film.[42] A continuous roll-to-roll coating apparatus has a moving metal strip or wire that is cleaned and coated with a graphene layer, after which the coated substrate is collected by a second roller.[43] The apparatus does not make use of thermal release adhesive tapes. Roll-to-roll processes are amenable for coating a substrate up to 1000 m in length with a graphene film and then transferring the graphene to another substrate.[44] For example, carbon was deposited onto a copper foil by chemical vapour deposition as the foil moved through a furnace at 1273 K and the graphene film was then grown by cooling the foil at a rate of about $10° \, s^{-1}$. A flexible thermal release tape was pressed onto the graphene layer by a roller and the foil was then immersed in an etching solution of ferric chloride to dissolve the copper. A flexible substrate was pressed onto the graphene and heat applied to two flexible tapes to transfer the graphene film from the original tape.

12.6 COMPOSITES AND NANOCOMPOSITES

Exfoliated or delaminated graphite can produce graphene nanoplatelets with thicknesses in the nanometre range depending on the number of graphite sheets that make up the nanoplatelets, typically 10 nm or less. Nanoplatelets have been used as fillers or reinforcements for composite materials. Sodium thiosulphate solution was added to an acidified dispersion of exfoliated graphite that resulted in precipitation of elemental sulphur according to the following equation:[45]

$$2HCl + Na_2S_2O_3 \rightarrow 2NaCl + S + SO_2 + H_2O \qquad (12.1)$$

Sulphur particles with a size in the range 30 nm to about 100 nm were distributed on the surface of graphene particles. The sulphur-graphene composite material has a potential application as a cathode material in a lithium-ion battery with improved cycling performance. In another approach graphite oxide powders with a particle size around 4.2 μm were placed in a sealed chamber containing oxygen at 0.5 M Pa.[46] Oxygen intercalated into the graphite oxide, the gas pressure was released and the graphite oxide exfoliated by heating at 623 K, after which a stable dispersion of graphite oxide nanoplatelets was produced in water. A water-soluble polymer such as polyvinyl alcohol was added to the nanoplatelet dispersion, after which the dispersion was dried. The drying process resulted in the polymer forming a coating on the nanoplatelets so that this structure represented a graphite oxide-reinforced polymer nanocomposite. It was also found that it is possible to exfoliate chalcogenide materials such as MoS_2, TaS_2 and WS_2 by intercalating oxygen under pressure, releasing the pressure and heating at 623 K. Incorporation of nanoplatelets into a polymer can generate dust that is a health hazard and in addition it can be difficult to obtain a uniform distribution of particles in the polymer.[47] In order to avoid dusts, a carbon nanomaterial such as graphite nanoplates were added to a liquid additive used for a polymer composite such as a low molecular weight polyolefin and pellets produced on a tablet press. The pellets were then blended with a polymer such as an acrylonitrile-butadiene-styrene mixture and subjected to injection moulding to produce test samples 10 cm x 5 cm x 0.2 cm in dimension. It can be difficult in general to obtain a uniform distribution of graphite fillers in a polymer matrix, because the mixture may be viscous, large quantities of organic solvent are required and heat-melt extrusion techniques are not appropriate for heat-sensitive materials.[48] In order to develop a more environmentally friendly process, an aqueous fluoropolymer latex dispersion was added to a colloidal dispersion of graphene oxide, after which coagulation of the components occurred.[48] The sediment was vacuum-dried to a powdered nanocomposite that was pressed and heated at 623 K to a composite film with thicknesses around 150 μm and filler content up to 1.5 weight %. The nanocomposite had potential applications as a protective coating towards HCl gas and the nanocomposite films were less permeable to HCl than un-filled polymer films.

Sporting equipment such as tennis rackets, golf clubs, skis, snowboards and running shoes may be designed to provide the user with a competitive advantage as well as to enhance the user's comfort and to protect them from injury.[49] This equipment is often made out of lightweight, thin materials, but a balance has to be achieved between losing the effectiveness of materials if they are too thin or weak and adding too much weight to the equipment, even if extra material increases properties such as wear resistance and impact resistance. Fibre-reinforced polymer composites have been used in tennis rackets so that the racket is lightweight and high-strength in the direction of tension of the fibres. Incorporation of graphene flakes into a polymer blend used as the matrix for a fibre-reinforced composite can potentially increase the compressive strength of the composite when it is moulded into the shape of the racket, possibly due to close interaction between graphene, fibre and matrix. The graphene content in the matrix is less than

10 weight % and examples of reinforcing fibres include silicon carbide and aramids such as KEVLAR. Graphene has also been considered as part of a fibre-reinforced composite for the soles of footwear to modify the stiffness of the sole and support given by the shoe to the user.[50] Chemical reduction of graphite oxide to graphene requires use of aggressive chemical reagents such as hydrazine. If the graphite oxide is a component of a polymer composite then this aggressive treatment may damage the polymer component. Exposure of composite films of single layer graphite and polystyrene to a single pulse from a Xenon flash light with optical energy in the range 0.1–2 J cm^{-2} reduced the graphite oxide component in the film to graphene.[51]

The use of graphene as a filler in composites is not restricted to polymer matrices. Table 12.2 shows the effect of added graphene on increasing the fracture toughness of silicon nitride for graphene concentrations up to 1.5 volume %.[52]

In situ growth of graphene took place in a sintered silicon carbide powder containing sintering aids such as alumina and yttrium oxide that was sintered in a die with a uniaxial pressure of 50 M Pa and at a maximum temperature of 2073 K using pulsed direct current spark plasma sintering.[53] Raman spectroscopy indicated the presence of graphene in the graphene/silicon carbide composites. Composites of graphene and metal salts were prepared by adding a salt solution such as $NiCl_2$ to a dispersion of graphite oxide, after which the mixture was dried and exposed to a focused beam of sunlight.[54] The latter decomposed and reduced the salt to produce a nanocomposite of metal or metal oxide nanoparticles dispersed on graphene particles several nanometres in diameter with a potential

Table 12.2 Effect of graphene on the physical and mechanical properties of silicon nitride composites.[52]

Material composition	Density g cm^{-3}	Per cent of theoretical density	Hardness G Pa	Toughness M Pa m$^{0.5}$
100 volume % Si_3N_4	3.223	100.0	22.3 ± 0.84	2.8 ± 0.12
99.98 volume % Si_3N_4	3.204	99.5	21.2 ± 0.34	2.7 ± 0.14
99.50 volume % Si_3N_4	3.198	99.7	19.3 ± 0.69	5.21 ± 1.00
99.00 volume % Si_3N_4	3.175	99.3	20.4 ± 0.37	5.8 ± 1.18
98.50 volume % Si_3N_4	3.175	99.6	15.7 ± 0.61	6.6 ± 1.31

application as anodes in batteries. Carbon nanotubes were mixed with a finely divided ceramic powder such as alumina that had been ball-milled to a median particle size of 700 nm. The mixture was subjected simultaneously to high pressure, typically about 35 M Pa and a temperature of 1873 K, which resulted[55] in conversion of the nanotubes to graphene ribbons within the sintered composite. The ribbons were between 1 to 100 nm in width, between 500 nm and 10 μm in length and between 0.4 nm and 2 nm thick. Composite powders of graphene oxide and manganese oxides for applications as electrode materials were prepared by first adding a suspension of manganese hydroxide to a dispersion of graphene oxide.[56] Then the mixture was subjected to spray pyrolysis in which it was converted to a spray that was passed into a furnace where drying and decomposition of salts took place. Graphene oxide is more hydrophilic than graphene nanoplatelets and more readily disperses in an aqueous environment.

12.7   DEVICES

Graphene has attracted much attention as a replacement for silicon-based semiconductor devices.[57] Its charge mobility, up to 2×10^5 cm^2 V^{-1} s^{-1} is about one hundred times faster than that of silicon and it has a current density around 10^8 A cm^{-2}, which is more than one hundred times greater than that of copper. These are desirable properties for the next generation of semiconductor devices. Graphene can be formed as a thin film on metal substrates such as Ni or Cu by using chemical vapour deposition or on a SiC substrate. In order to use a thin graphene layer as a semiconductor device it is necessary to grow the film on an insulating layer and the ability to remove and transfer graphene films from one surface to another surface is an important process. One way in which growth of a graphene sheet on an insulator has been achieved is to deposit an insulating layer such as a metal oxide or metal nitride by, for example, molecular beam epitaxy or chemical vapour deposition. A thin layer of a metal carbide such as silicon carbide or tungsten carbide is then grown to a depth of around 10 atomic layers on top of the insulating layer.[58] The substrate is then subjected to pulse annealing in a high vacuum that raises the temperature of the metal carbide layer to between 1173 K and 1773 K. Metal ions migrate towards the interface with the vacuum along with carbon atoms that form a graphene layer with the geometric size of the underlying wafer and which can be subjected to standard lithographic patterning processes.

Graphene can function as a channel in a field effect transistor (FET) and Figure 12.5 outlines a stage in the production of the transistor.[59] A key feature is the use of a sacrificial seed material, nickel, in the form of a layer up to 500 nm, that is deposited by sputtering onto the walls of a mandrel. The nickel can be patterned by photolithographic masking and etching.

The mandrels are fabricated by depositing a layer of SiO_2 on the substrate and then subjected to semiconductor processing techniques such as masking, patterning and etching. A single layer of graphene is deposited as a layer onto the nickel seed by CVD and the nickel is then removed by wet etching in which hydrochloric acid is introduced through the vent hole shown in Figure 12.5.

Figure 12.6 illustrates another design for a field effect transistor and a key feature of this design is that the graphene layer could be grown on one substrate by CVD and then transferred by an adhesive tape onto the structure shown in Figure 12.6.[60]

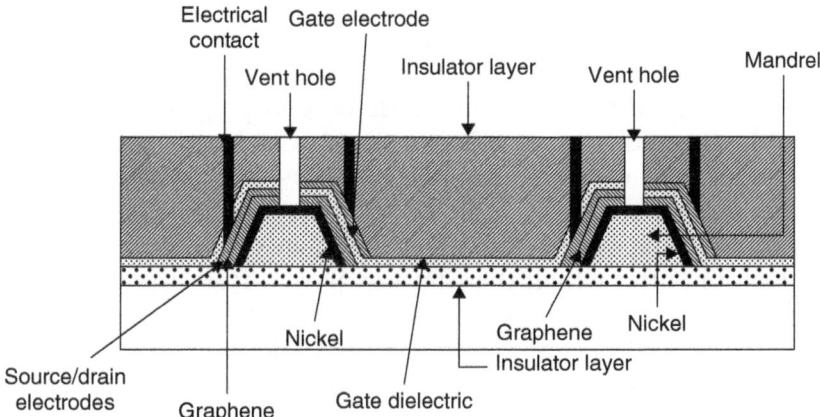

Figure 12.5   Schematic diagram of a graphene-based field effect transistor based on a sacrificial nickel layer.[59]

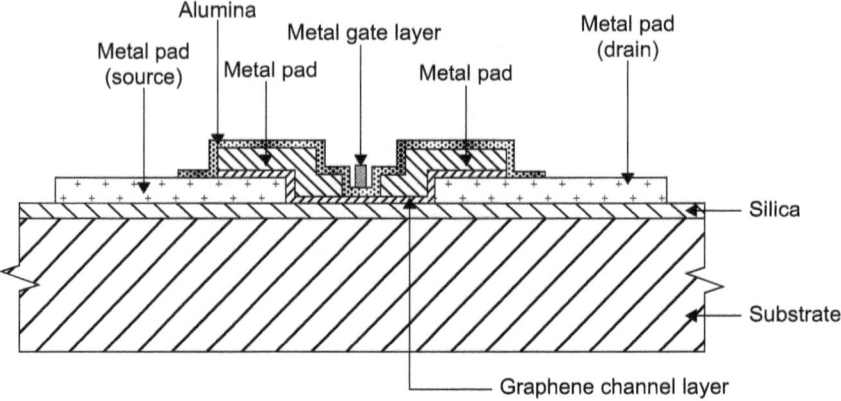

Figure 12.6   Schematic diagram of a graphene-based field effect transistor based on a transferable graphene layer.[60]

Further examples of the use of graphene in semiconductor devices are discussed in references 61–66.

12.8   BATTERIES AND SUPERCAPACITORS

The increasing use of mobile devices such as smartphones and tablet computers as well as the development of electric vehicles has led to a demand for rechargeable batteries. Lithium-ion batteries are used widely in portable devices and consist of a carbon anode and a lithium cobalt oxide ($LiCoO_2$) as the cathode.[67] During charging/discharging cycles, lithium ions are transported through the electrolyte and form intercalation compounds with the carbon anode on charging or are inserted into the layered lithium cobalt oxide on discharging. Graphene, as a source of carbon, has attracted considerable attention because of its potential use in electrode materials. For example, graphene films were grown on nickel plates by chemical vapour deposition, after which the film was peeled from the plates and immersed in dilute mineral acid to dissolve Ni. The graphene film had a multi-layered structure that allowed intercalation of lithium ions when evaluated for use as an anode. Supercapacitors, also referred to as ultracapacitors, have the potential to provide higher energy densities more quickly than batteries and electrochemical supercapacitors include capacitors based on ion insertion compounds.[68] Graphene has been considered as one of three components for a supercapacitor electrode. Hence the electrode contained a material such as graphene that acted as a double layer supercapacitor and was a highly structured carbon, a less-structured form of carbon such as activated carbon or carbon black while the third component consisted of oxide nanoparticles. Activated carbon was considered to improve the electrical contact between the oxide nanoparticles and graphene. Graphene nanoplatelets have been used to reinforce a matrix used as a cathode active material.[69] As an example, particles of lithium vanadium oxide, γ-LiV_2O_5, were made by a hydrothermal process and had a length between 0.3 and 3 µm and a diameter of approximately 30–50 nm. These nanorods were mixed with graphene nanoplatelets, a phenolic resin and acetone and subjected to spray pyrolysis to produce nanocomposite particles. The particles were cured at 473 K and carbonised at 873 K to yield a composition of about 88 weight % γ-LiV_2O_5 nanorods, 5 weight % nanoplatelets and 7 weight % carbon. The graphene increased the strength and fracture toughness of the carbon matrix.

12.9 SUMMARY

Graphene, a two-dimensional sheet of carbon that is one atomic layer thick, has received much publicity since its isolation in 2004 and has generated considerable excitement within scientific circles. This is because its electronic and mechanical properties, such as electrical conductivity, electron mobility, mechanical strength and optical transparency have many potential applications including as replacements for silicon-based semiconductor devices, fillers for composites and components of electrodes for lithium-ion batteries. Materials chemistry has a very important role in the development of methods for the preparation of graphene.

REFERENCES

1. J. G. Bednorz and K. A. Muller, Possible high-Tc superconductivity in the Ba-La-Cu-O system, *Zeitschrift fur Physik B*, 1986, **64**, 189–193.

2. A. Geim, K. Novoselov, R. Gorbachev and L. Ponomarenko, Structures and methods relating to graphene, *United States Patent Application*, 2014/0008611, 2014.

3. B. Z. Jang, L. Yang, S.-C. Wong and Y. Bai, Process for producing nano-scaled graphene plates, *United States Patent Application*, 2005/0271574, 2005.

4. B. A. Anderson and E. J. Nowak, Graphene-based transistor, *United States Patent*, 7 732 859, 2010.

5. J. Liu, I. A. Aksay, D. Choi, D. Wang and Z. Yang, Nanocomposite of graphene and metal oxide materials, *United States Patent Application*, 2014/0030181, 2014.

6. M. Zhou, J. Pan and Y. Wang, Composite material of carbon-coated graphene oxide, preparation method and application thereof, *United States Patent Application*, 2013/0344393, 2013.

7. M. Rafailovich, R. Isseroff and P. M. Das, Chemical synthesis for graphene sheets greater than 1 μm in length, *United States Patent Application*, 2011/0281035, 2011.

8. J.-Y. Choi, H.-J. Shin and S.-M. Yoon, Graphene sheet and process of preparing the same, *United States Patent*, 8 075 864, 2011.

9. S. Eigler and A. Hirsch, Preparation method for graphene oxide suitable for graphene production, *European Patent Application*, 2 639 201 A, 2013.

10. Y. Miyamoto, Structure of graphene oxide, the method of fabrication of the structure, the method of fabricating field-effect transistor using the structure, *United States Patent Application*, 2014/0017440, 2014.

11. H. Lee and P. Cul, Novel graphene oxide reducing agent and method for preparing reduced graphene oxide using the same, *United States Patent Application*, 2013/0323159, 2013.

12. Z. Liu, X. Zhou, Z. Qin and C. Tang, Method for preparing graphene, *United States Patent Application*, 2014/0037531, 2014.

13. G. Krishnaiah and V. Varma, Process for the preparation of graphite oxide and graphene sheets, *United States Patent Application*, 2012/0128570, 2012.

14. Y.-S. Wu, C.-Y. Hsieh, C.-S. Peng, J.-R. Chen, J.-M. Lin and G.-W. Lin, Method for the preparation of graphene, *United States Patent Application*, 2013/0197256, 2013.

15. W.-K. Choi, D. H. Park, B. W. Kwon and D. I. Son, The method for producing graphene by chemical exfoliation, International Patent Application, 2012/165753, 2012.

16. X. Chen, Z. Niu and J. Ma, Method of preparing reduced graphene oxide foam, *United States Patent Application*, 2013/0314844, 2013.

17. A. Penicaud and C. Valles, Graphene solutions, *United States Patent Application*, 2011/0130494, 2011.

18. C. Howard, N. Skipper, M. Shaffer and E. Milner, Dispersion method, *International Patent Application*, 2013/001266, 2013.

19. Y. J. Yun and K. Song, Method of manufacturing a graphene fiber, *United States Patent Application*, 2013/0272950, 2013.

20. S. J. Kim, M. K. Shin and S. H. Kim, Graphene fiber and method for manufacturing same, *European Patent Application*, 2 687 626 A, 2014.

21. J.-Y. Choi, Graphene sheet comprising an intercalation compound and process of preparing the same, *United States Patent*, 8 227 685, 2012.

22. M. Gharib, A. I. Aria and A. W. Gani, Method for producing graphene oxide with tunable gap, *United States Patent*, 8 609 458, 2013.

23. K. Coleman, Process for producing graphene, *International Patent Application*, 2012/172338, 2012.

24. G. Dong and R. Van Rijn, Thin film formation, *United Kingdom Patent Application*, 2 498 944 A, 2013.

25. G. Pan, Production of graphene, *United Kingdom Patent Application*, 2 502 533 A, 2013.

26. P. Papakonstantinou and N. Shang, Process for the preparation of graphene, *United States Patent Application*, 2014/0044968, 2014.

27. R. M. Frazier, D. T. Daly, S. K. Spear and R. D. Rogers, Exfoliation of graphite using ionic liquids, United States Patent Application, 2011/0319554, 2011.

28. J. E. Kim and T. Y. Kim, Graphene dispersion and graphene-ionic liquid polymer compound, *European Patent Application*, 2 518 103 A, 2012.

29. R. A. W. Dryfe and I. A. Kinloch, Production of graphene, *United States Patent Application*, 2014/0061059, 2014.

30. Y.-T. Hsieh, K.-P. Huang and P. Lin, Methods of forming graphene, *United States Patent Application*, 2013/0164208, 2013.

31. A. Geim, R. Raveendran-Nair and K. Novoselov, Functionalized graphene and methods of manufacturing the same, *United States Patent Application*, 2011/0303121, 2011.

32. M. Zhou, D. Liu and Y. Wang, Fluorographene and preparation method thereof, *European Patent Application*, 2 657 188 A, 2013.

33. M. G. Schwab, K. Mullen, X. Feng and K. Parvez, Graphene containing nitrogen and optionally iron and/or cobalt, *International Patent Application*, 2014/012600, 2014.

34. J. A. Bowers, R. A. Hyde, M. Y. Ishikawa, J. T. Kare, C. T. Tegreene, T. Toyokuni and R. N. Zare, Doped graphene electronic materials, *United States Patent*, 8 426 842, 2013.

35. I. A. Aksay, D. L. Milius, S. Korkut and R. K. Prud'Homme, Functionalized graphene sheets having high carbon to oxygen ratios, *United States Patent Application*, 2011/0114897, 2011.

36. H. Dai, X. Li and X. Sun, Pristine and functionalized graphene materials, *United States Patent Application*, 2010/0028681, 2010.

37. S. Hellstrom, M. Vosgueritchian, Z. Bao and M.-G. Kim, Doping of carbon-based structures for electrodes, United States Patent Application, 2013/0330559, 2013.

38. N.-W. Pu, C. A. Wang, Y. Sung and M.-D. Ger, Method for manufacturing graphene, *United States Patent*, 8 414 799, 2013.

39. H. Todoriki, Y. Takemura and K. Nomoto, Graphite oxide, graphene oxide or graphene, electric device using the same and method of manufacturing the same, and electrodialysis apparatus, *United States Patent Application*, 2013/0183226, 2013.

40. W. Strupinski, Method of graphene manufacturing, *International Patent Application*, 2011/155858, 2011.

41. S.-H. Cho and D.-K. Won, Method of transferring graphene, *United States Patent*, 8 419 880, 2013.

42. D.-H. Na, J.-H. Yoon, Y.-I. Song and D.-K. Won, Method and apparatus for transferring graphene, *United States Patent Application*, 2013/0233480, 2013.

43. B. H. Hong, Y. J. Kim, J. Choi, H. K. Kim, J. Kang and S. K. Bae, Graphene roll-to-roll coating apparatus and graphene roll-to-roll coating method using the same, *United States Patent Application*, 2011/0195207, 2011.

44. B. H. Hong, J. Ahn, S. Bae and H. K. Kim, Roll-to-roll transfer method of graphene, graphene roll produced by the method, and roll-to-roll transfer equipment for graphene, *United States Patent Application*, 2012/0258311, 2012.

45. L. Wang, X.-M. He, J.-J. Li, J.-W. Guo, W.-T. Sun and J.-G. Ren, Method for making sulphur-graphene composite material, *United States Patent*, 8 609 183, 2013.

46. B. Z. Jang, A. Zhamu and J. Guo, Process for producing nano-scaled platelets and nanocomposites, *United States Patent Application* 2008/0048152, 2008.

47. S. H. Ryu, K. Ju, N. S. Choi, M. W. Jung, Y. H. Song and Y. C. Jang, Method for preparing carbon nanomaterial/polymer composites, *United States Patent Application*, 2013/0207052, 2013.

48. F. Gauthy, A. Sanguineti, F. D'Aprile and P. D'Orazio, Process for producing graphene-polymer nanocomposites, *International Patent Application*, 2013/127712, 2013.

49. H. Lammer, Sporting goods with graphene material, *United States Patent Application*, 2013/0090193, 2013.

50. T. L. Torrance and P. Majure, Flexible shoe sole, *United States Patent Application*, 2014/0068880, 2014.

51. J. Huang, L. Cote and R. C. Silva, Method of flash reduction and patterning of graphite oxide and its polymer composites, *United States Patent Application*, 2010/0221508, 2010.

52. E. L. Corral, L. S. Walker, V. R. Marotto, M. A. Rafice and N. Koratkar, Graphene-reinforced ceramic composites and uses thereof, *United States Patent Application*, 2013/0184143, 2013.

53. P. Miranzo, C. Ocal, M. I. Osendi, M. Belmonte, C. Ramirez, B. Roman-Manso, H. R. Guitierrez and M. Terrones, Process for production of graphene/silicon carbide ceramic composites, *International Patent Application*, 2014/047283, 2014.

54. R. Sundara, E. Varrla and J. A. Sasidharannair Sasikaladevi, Graphene composites with dispersed metal or metal oxide, *United States Patent Application*, 2014/0054490, 2014.

55. D. C. Ogrin, K. R. Kissell, J. C. Falkner, K. L. Lundberg and J. R. Tidrow, Ceramic matrix composite articles comprising graphene nanoribbons-like material and their manufacturing method using carbon nanotubes, *International Patent Application*, 2011/086382, 2011.

56. K. K. Konstantinov, Spray pyrolysis method for in situ production of graphene oxide based composites, *International Patent Application*, 2012/155196, 2012.

57. W. Xianyu, C.-Y. Moon, J.-Y. Lee and C.-S. Lee, Graphene devices and methods for manufacturing the same, *United States Patent Application*, 2013/0161587, 2013.

58. C. Ma, Formation of graphene wafers on silicon substrates, *United States Patent*, 7 947 581, 2011.

59. J. W. Adkisson, T. J. Dunbar, J. P. Gambino and M. J. Leitch, Graphene field effect transistor, *International Patent Application*, 2013/089938, 2013.

60. J.-H. Lee, T.-H. Jeon, Y.-S. Kim, C.-S. Lee and Y.-S. Jung, Graphene device and method of manufacturing the same, *United States Patent Application*, 2014/0061590, 2014.

61. A. Chen and Z. Krivokapic, Device and process of forming device with device structure formed in trench and graphene layer formed thereover, *United States Patent*, 7 858 989, 2010.

62. H. Numata, S. Toguchi and H. Endoh, Field effect transistor and circuit device, *United States Patent Application*, 2011/0114914, 2011.

63. C. D. Dimitrakopoulos, A. D. Franklin and J. T. Smith, Transport conduits for contacts to graphene, *United States Patent Application*, 2013/0337620, 2013.

64. J. B. Chang, W. E. Haensch, F. Liu and Z. Liu, Graphene devices and silicon field effect transistors in 3D hybrid integrated circuits, *United States Patent*, 8 587 067, 2013.

65. I. Zaliznyak, A. Tsvelik and D. Kharzeev, Nanodevices for spintronics and methods of using same, *United States Patent*, 8 378 329, 2013.

66. X. Duan, Y. Huang, L. Liao and J. Bai, High performance graphene transistors and fabrication processes thereof, *United States Patent Application*, 2014/0077161, 2014.

67. C.-M. Sung, Li-ion battery, *United States Patent Application*, 2012/0052387, 2012.

68. G. Gruner and I. O'Connor, Charge storage device for increasing energy and power density, *United States Patent*, 8 520 365, 2013.

69. A. Zhamu, B. Z. Jang and J. Shi, Nano graphene reinforced nanocomposite particles for lithium battery electrodes, *United States Patent*, 8 580 432, 2013.

CHAPTER 13

Hydrogels

13.1 INTRODUCTION

The variety of consumer goods available for purchase by the general public has increased markedly in recent years. For example, smartphones and tablet computers are widely used in everyday life by millions of people. Most people will be unaware of the range of materials and their discrete properties, magnetic, electrical, optical or mechanical that are used in the construction of, for example, smartphones. The appeal of consumer goods is made up in part by their design, rather than the technical specification determined by the materials of construction that are often hidden from view. Indeed, certain models of smartphone or tablet computer that are desirable objects for purchase are often described as being 'cool'. The latter word is difficult to define but its use in a recent court case highlights its meaning. An infringement action case in the United Kingdom involved Samsung Electronics (UK) Limited and Apple Inc. in the area of tablet computers.[1] In a summarising statement, His Honour Judge Birss QC gave the informed user's overall impression of the Samsung Galaxy Tablets, which were referred to in the case: 'From the front they belong to the family which includes the Apple design; but the Samsung products are very thin, almost insubstantial members of that family with unusual details on the

Exploring Materials through Patent Information
By David Segal
© David Segal, 2015
Published by the Royal Society of Chemistry, www.rsc.org

back. They do not have the same understated and extreme simplicity which is expressed by the Apple design. They are not as cool. The overall impression produced is different'. It was concluded that the Samsung tablets did not infringe Apple's registered design. Properties of materials are very important in other consumer goods but as in the case of smartphones and tablet computers the materials are often hidden from the user. Many products are purchased for their functional properties, for example, detergents, soaps and disinfectants, although the extent of materials development that is used in their production may not be appreciated. Nappies (diapers) and contact lenses are widely used, but a majority of shoppers are unlikely to be aware that highly absorbent polymeric materials known as hydrogels are used in these products. Hydrogels are described in this chapter, together with potential applications, as viewed from the perspective of patent literature.

13.2 HYDROGELS

13.2.1 Definition

Aerogels were described in Chapter 9 as an example of a class of porous materials with a very low density. Hydrogels are another class of porous materials. They are three-dimensional networks of hydrophilic homopolymers or copolymers that can be swollen by absorption and retention of water.[2,3] They exhibit both liquid properties, because the major constituent is water and solid properties because of cross-linking during polymerisation. While the polymers themselves are soluble in water, the hydrogel is insoluble due to the presence of covalent, ionic or physical crosslinks. Hydrogels can be classified as amorphous, semi-crystalline, hydrogen-bonded structures, supermolecular structures or hydrocolloidal aggregates.[3] Porosity, pore size, the nature of the gel polymer, molecular weight of the polymer and crosslinking density are parameters that affect the physical properties (*e.g.* elasticity, compressive strength) of the hydrogel. Hydrogels have been described as superabsorbent and super-porous and the latter, which can have pores with diameters in the micrometre to millimetre range can absorb many times their weight of aqueous fluids in a few seconds.[4,5] The degree of crosslinking affects the absorbent capacity and gel strength of a superabsorbent.[6] Hydrogels can contain as much as 99.9 weight % water without losing their structural integrity.[7]

13.2.2 Preparation and Properties

The synthesis of hydrogels is described in this section, with some examples from the patent literature. Further examples are described in section 13.3 relating to specific applications for hydrogels. The polymer backbone of hydrogels is formed of hydrophilic monomer units that can be neutral or ionic.[8] Examples of neutral and hydrophilic monomer units include vinyl alcohol, ethylene oxide, vinylpyrrolidone, hyroxyalkyl methacrylates such as hydroxyethylmethacrylates and *N*-vinylacetamide. Examples of anionic monomer units are fumaric acid, vinylsulphonic acid, maleic acid and 4-vinylbenzenesulphonic acid. Examples of cationic monomer units include diallyldimethylammonium chloride and trimethylammonium propylmethacrylamide. Superabsorbents have been obtained by polymerisation of unsaturated acids, for example, acrylic acid in the presence of unsaturated olefinic compounds.[8] It is important to note that hydrogels are often dried, ground and classified before use in, for example, articles for personal hygiene such as nappies and sanitary towels. Highly porous hydrogels with a low density are analogous to aerogels. Table 13.1 shows the properties of a low-density hydrogel prepared from sodium alginate.[9]

The hydrogel was prepared by first adding calcium chloride solution to a solution of sodium alginate. Calcium alginate was formed as pellets that were crosslinked with 2,4-tolylene diisocyanate in toluene in the presence of a base catalyst, triethylamine. Note that calcium alginate, which was not crosslinked had an apparent bulk density of 0.243 g cm^{-3} and a surface area of 20.7 m^2 g^{-1}. Silicone-containing hydrogel, for

Table 13.1 Properties of crosslinked alginate.[9]

Apparent bulk density g cm^{-3}	0.042
Surface area m^2 g^{-1}	200
Pore volume cm^3 g^{-1}	2.92
Average pore diameter nm	52

example, 1,3-bis(4-methacryloxybutyl) tetramethyldisiloxane for use as a contact lens has been prepared by reaction of an acrylic-capped polysiloxane such as 1,3-bis(4-hydroxybutyl)-tetramethyldisiloxane with methacryloyl chloride.[10] Polyurethane hydrogels have been prepared by a free-radical cross-linking process.[11] For example, a hydrophilic polyurethane having an unsaturated olefinic group, either acrylate or methacrylate (*e.g.* urethane acrylate) is reacted with a water-soluble oxidising agent and water-soluble reducing agent in which the polyurethane is based on a polyalkylene oxide. Here, the hydrogel was obtained by mixing the urethane acrylate with a mixture of ammonium peroxodisulphate, ascorbic acid and ferrous chloride.

Hydrogels are formed from natural, as well as synthetic polymers. Cyclodextrins are formed enzymatically from starch by the action of cyclodextrin glycosyltransferase produced from microorganisms and are cyclic molecules containing six, seven or eight alpha-D-(+)-glucopyranose rings bonded together by 1,4 linkages.[12] A partially polymerised prepolymer of polyurethane was produced through the reaction of 2,4-toluene diisocyanate and 2,6-toluene diisocyanate, with a polyethylene oxide and polypropylene oxide in which the prepolymer has terminal isocyanate groups. A hydrogel was produced by allowing the prepolymer to react with and crosslink the cyclodextrin and the reaction could take on a substrate such as natural fibres including cotton and pressure-sensitive adhesive dressings for personal care products including nappies. A further example of a hydrogel produced from a natural polymer relates to the crosslinking reaction between hyaluronic acid (also known as hyaluronan), a naturally occurring water-soluble polysaccharide and proteins such as elastin for use as injectable dermal fillers.[13] Crosslinking agents included 1-ethyl-3-(3-dimethylaminopropyl) carbodiimide and N-hydroxysuccinimide. Hydrogels have also been formed between hyaluronic acid the natural product chitosan.[7] Superporous hydrogels can be prepared from hydrophilic monomers, including acrylic acid and its salts, acrylamide, the potassium salt of sulphopropyl acrylate, hydroxyethyl acrylate and hydroxyethyl methacrylate.[14] The ingredients and their role in the synthesis of a superporous hydrogel are shown in Table 13.2.[14]

Biodegradable hydrogels have been prepared.[15,16] For example, poly(dichloro phosphazene) dissolved in tetrahydrofuran

Table 13.2   Ingredients and their role in the preparation
of a superporous hydrogel.[14]

Ingredient	Role
Hydroxyethylacrylate	Monomer
Poly(ethylene glycol) diacrylate	Crosslinker
De-ionised water	Diluent/solvent
Pluronic® F127	Surfactant
Glacial acetic acid	Foaming aid
Sulphopropylacrylate (potassium salt)	Co-monomer
Carboxymethylcellulose (sodium salt)	Polysaccharide
Tetramethylethylenediamine	Reductant
Ammonium persulphate	Oxidant
Sodium bicarbonate	Blowing agent

was reacted with phenylalanine ethyl-ester hydrochloride and triethylamine at room temperature.[15] A solution of triethylamine and aminoethoxy polyethylene glycol in tetrahydrofuran was then mixed with this solution and left for 48 hours at room temperature. The combined solution was then added to lysine ethyl-ester hydrochloride and triethylamine dissolved in tetrahydrofuran and reacted at room temperature for 48 hours. A hydrogel, poly[(phenylalanine ethyl-ester) (aminoethoxy polyethyleneglycol 350) (lysine ethyl-ester) phosphazene)], was obtained by precipitation with hexane. An application for these biodegradable hydrogels is for the delivery of drugs. Other phosphazene-based hydrogels have been described.[16] Biodegradable hydrogels derived from macromolecules, such as dextrans containing polymerisable side groups have been prepared.[17] For example, dextran was derivatised with hydroxyethyl methacrylate by coupling hydroxyethyl methacrylate to dextran by using a carbonyldiimidazole activator. Hydrogels were obtained by a free radical polymerisation of aqueous solutions of methacrylated dextran at room temperature by using potassium peroxydisulphate as the source of free radicals.

Hydrogels have been prepared from fluorinated polymers.[3,18] For example, from hydrophilic polymers containing perfluorocyclobutane crosslinking segments that link polymer chains together.[3] Another approach to fluorinated hydrogels involves the use of fluorinated supramolecular polymers that contain at least 5 weight % of covalently bonded fluorine atoms whereby the ratio of hydrophilic and hydrophobic components in the polymer are carefully balanced.[18] While hydrogels are often obtained as dry powders, hydrogel foams and hydrogel fibres have been prepared.[19,20] For example, poly(acrylic acid) foams were prepared by using *N,N'*-methylene-bis-acrylamide as a crosslinking agent at a 1 weight per volume % concentration in acrylic acid.[19] Ammonium persulphate was used to instigate the free radical reaction and sodium bicarbonate was added to the reactants so that carbon dioxide released in the acidic conditions in solution caused rapid foaming. Polyacrylic acid hydrogel foams reached 70% of their equilibrium swelling within 30 minutes, while conventional polyacrylic acid hydrogels took more than 3 days to reach 70% of the equilibrium swelling. Alginate is a biodegradable polymer obtained from seaweed and is a linear polysaccharide copolymer.[20] Hydrogel fibres of barium alginate were formed spontaneously on extruding a sodium alginate solution through a needle into a bath of barium chloride solution.[20] The water content of the fibres reached up to 99.5% of the weight of the fibre.

Concise reviews on the preparation of hydrogels are described in the patent literature and the interested reader is recommended to peruse the descriptions given in references 3, 21 and 22. In Chapter 8, an overview of phase separation in block copolymers was given in terms of changes in free energy and the miscibility of polymers. There are analogies between phase separation of block copolymers and the formation of hydrogels, as the latter also involves mixing of polymers and changes in free energy.

The properties of hydrogels give rise to a wide range of applications. These properties include transparency, permeability to oxygen, biocompatibility, the ability to absorb almost 100% of their weight as water while maintaining structural integrity, insolubility in aqueous media and examples of these applications are described in section 13.3.

13.3 APPLICATIONS FOR HYDROGELS

13.3.1 Contact Lenses

Hydrogels are used as contact lenses. There are two types of contact lenses: 'soft' and 'hard'. Soft lenses are mainly hydrogels

Table 13.3 Mixture of components for producing a soft contact lens.[26]

Component	Weight %
2-hydroxy-2-methyl-1-phenyl-propan-1-one	1
N,N-dimethylacrylamide	35
α,ω-bismethacryloxypropyl polydimethylsiloxane	36.5
Tris(trimethylsiloxy) silylpropyl methacrylate	27.5

derived from hydrophilic monomers or polymers that have been crosslinked. They contain around 38 weight % of water. Hard lenses had been prepared from polymethylmethacrylate but because these systems lacked sufficient oxygen permeability to provide oxygen to the cornea hard oxygen-permeable materials were introduced. Fluoromethacrylates[23] have been used for hard lenses to improve hardness, refractive index and wettability. Soft lenses are made by casting into a mould using two mould halves.[24,25] A pre-polymer mixture is poured into the moulds, which are pressed together, after which a curing step occurs, for example, by application of ultraviolet light that instigates a polymerisation reaction. Table 13.3 shows a blend of components used to produce soft contact lenses.[26]

The blend shown in Table 13.3 was diluted with tert-butanol. The resulting clear, homogeneous solution was poured into moulds and irradiated with ultraviolet light. Clear lenses were obtained. Polymeric hydrogel contact lenses containing desferrioxamine as an adsorbed species were prepared.[27] Desferrioxamine, which was capable of leaching out from the lens into the lachrymal fluid showed potential as an inhibitor for bacterial growth. Coloured hydrogel contact lenses have an aesthetic appeal.[28] Photocurable inks are printed onto polypropylene lens moulds and the latter were filled with a silicone-based fluid that was cured by irradiation with ultraviolet light. Cured ink patterns were formed on the surfaces of the lenses and printed ink could also be sandwiched between the two halves of a hydrogel lens. The oxygen permeability of soft contact lenses can be increased by adding silicone-containing monomers to the hydrogel formulations.[29] Examples of these monomers include alkyl-terminated polydimethylsiloxanes, methacryloxypropylpentamethylsdisiloxane and methacryloxypropyltris (trimethylsiloxy)silane. Silicone hydrogels had a low Young's modulus, a desirable property as lenses that were not rigid, that is, not stiff and quickly returned to their original shape when deformed. Silicone monomers were gelled on exposure to ultraviolet light or cured thermally at temperatures up to 373 K.

13.3.2 Cosmetics

γ-polyglutamic acid and its salts can be produced in a fermentation process with *Bacillus licheniformis* by using L-glutamic acid and glucose as feedstocks.[30] The chemical structure of γ-polyglutamic acid is shown in Figure 13.1.

Cross-linked hydrogels of γ-polyglutamic acid can be produced by irradiation with electron beams, gamma rays or with polyfunctional chemical cross-linking agents, for example, diglycerol polyglycidyl ether. The hydrogels are used as moisturisers in cosmetic and personal care products for skin-care and hair-care at concentrations in the range 0.005 to 5 weight % based on the final product. Figure 13.2 shows the chemical structure of 2-deoxy-2-{[N-(naphthalen-2-ylacetyl)-D-phenylalanyl]amino}-D-glucopyranose and hydrogels of this compound are obtained on heating an aqueous solution at 323–358 K.[31] These hydrogels could be used as components of moisturisers for skin on the face, neck and forearms and other regions of skin on the body.

A further example of the use of hydrogels in cosmetic and personal care products relates to the addition of hydrogels derived from carrageenan to soap bars.[32] These hydrogel fillers allow the amount of fatty material in the soap to be reduced, while fragrances can be incorporated into the hydrogel matrix.

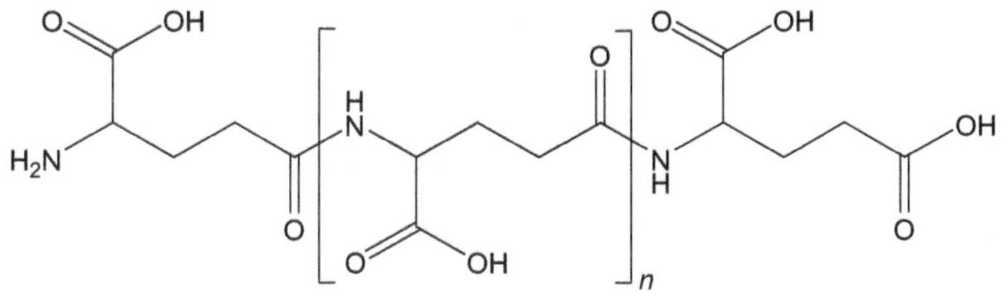

Figure 13.1 Chemical structure of γ-polglutamic acid.[30]

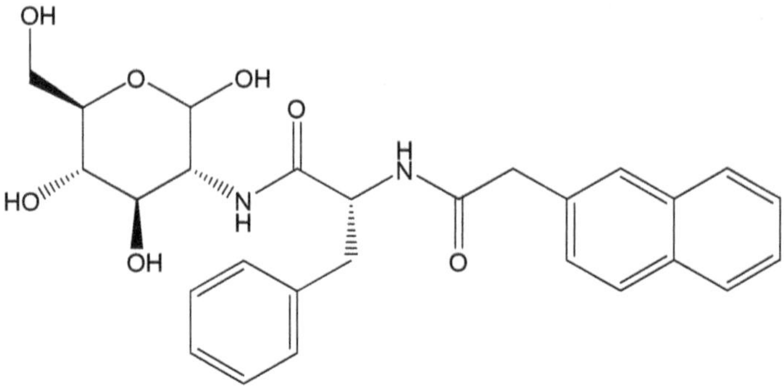

Figure 13.2 Chemical precursor for a hydrogel used in a moisturiser.[31]

13.3.3 Wound Dressings

Their ability to absorb and retain large quantities of fluid relative to their weight make hydrogels candidates for wound dressings. It should be noted that structures used to contain hydrogels for dressings can be applicable to other articles such as sanitary towels and nappies.[33] As an example, isophorone diisocyanate prepolymer was mixed with polyethylene glycol and, separately, propylene glycol was mixed with polyetherdiamine.[34] The two mixtures were combined and cast into a mould, when gelation occurred after 90 minutes at room temperature. An antimicrobial agent, silver sulphadiazine could be incorporated into the reactants before gelation. The blend of reactants could be cast into hydrogel strips that were strong enough to be applied to wounds without the use of backing material. Additives for slow release can be added to the hydrogel blends before casting, for example, zinc in the form of zinc lactate for use as an anti-oxidant and anti-inflammatory compound and glucose to help the growth of glucose-containing biopolymers.[35] The hydrogel, for example, one prepared from bovine gelatin cross-linked with glutaraldehyde or formaldehyde can be sandwiched between a perforated polyethylene film in contact with the wound and an absorbent layer.[36] The hydrogel can be a polyurethane gel or a hydroxyethyl cellulose gel, for example. The hydrogel layer helps to keep the wound moist. A layered material consisting of a hydrophilic polyurethane foam layer and a hydrogel layer formed from a polyurethane has been used as a wound dressing.[37]

The thickness of the foam layer is between 0.15 and 0.5 cm, while the hydrogel layer had a thickness between 0.2 and 0.5 cm. The hydrophilic foam has a more rapid uptake of fluid than the hydrogel layer alone. It is important for hydrogel wound dressings to remain attached to the wound during day-to-day activities, and hydrogel adhesives have been developed specifically for use with wound dressings.[38]

13.3.4 Hydrogel Adhesives

Medical procedures sometimes require measurement of electrical signals from a patient's body and devices such as electrodes are temporarily attached to the patient. It is important for the attachment to remain in place on both wet and dry skin and to be removed painlessly from the patient. Hydrogel adhesives or bioadhesives have been developed for medical diagnostics. For example, a precursor composition of acrylic acid, diisopropanolamine, a photoinitiator and a cross-linking agent was applied to the medical device and cured by exposure to ultraviolet light.[39] The coated device could be attached to both wet and dry skin. In another approach, glycerol was added to a solution of sodium acrylate.[40] A polyoxymethylene triacrylate crosslinking agent was added to the solution of monomers that were cured by exposure to ultraviolet light. The hydrogel adhesive could be attached to skin and could also be used in personal care products such as absorbent articles. A further example on bioadhesives is given in reference 41.

13.3.5 Agriculture

The ability to absorb and retain water makes hydrogels ideal candidates as components in soils in arid environments.[42] Hydrogels were dried, ground and screened to superabsorbent powders with preferred particle sizes between 106 μm and 850 μm before adding to the soil at a hydrogel content less than 1 weight %. Irrigation trials were carried out to determine the optimum amount of water to apply to the soils based on the so-called suction pressure in the soil. Superabsorbent powders based on starch graft copolymers have been used as a root dip for plants.[43] As an example, a superabsorbent powder based on the potassium salt of starch-g-poly (2-propenamide-co-2-propenoic acid) was dispersed in water until it was fully hydrolysed. The roots of a plant were dipped into the hydrogel and, once planted, the hydrogel could provide moisture to the root system. Biodegradable hydrogels have been used as protective coatings for seeds.[44] A gelatin-based hydrogel formulation based on a naturally derived hydrophilic protein such as a porcine protein and a sulphated polysaccharide was used to coat seeds by dipping the latter into the hydrogel. The hydrogel coating reduced the need for repeated watering of the seeds and acts as a reservoir of water for the seeds.

13.3.6 Drug Delivery

About 80% of all drug dosages on the market are taken in oral form.[45] Orally administered pharmaceutical agents must be transported to the stomach and small intestine for absorption across the gastrointestinal mucosal membranes into the blood. The efficiency of absorption can be low because of metabolism within the gastrointestinal tract. Hence, improved methods for drug delivery are required. One approach is to use small tablets with a volume about 10^{-2} cm^3 that contains sufentanil and a hydrogel-forming compound that can act as a bioadhesive, causing adherence to the oral mucosa of the patient. Another approach is to form a codrug, that is, a compound in which two drugs are chemically linked, for example, triamcinolone acetonide and 5-fluorouracil.[46] The chemical linkage is broken under physiological conditions and the composition hydrates to form a hydrogel. Transdermal methods of drug delivery in which the active ingredient passes through the skin are under development. For example, a plaster in which a hydrogel layer contains diclofenac and a heparin or heparinoid is placed in contact with skin to promote transfer of the drugs into the patient.[47]

13.3.7 Additional Applications

Hydrogels are not a laboratory curiosity. It has been stated that the market for hydrogels is over one billion pounds per year in the US and about 2.5 times that globally, with a growth rate of approximately 3% per year.[48] Additional examples of

Table **13.4** Applications for hydrogels from the patent literature.

Application	Reference
Stereolithography	49
Screen printing biosensors	50
Surgery (cartilage tissue)	51
Surgery (implants)	52
Scaffold for tissue engineering	53
Bone regeneration	54
Downhole sealant	55
Groundwater barrier	56
Air freshener	57
Fabric conditioner	58
Conducting polymers	59
Stem cells	60
Biomedical articles	61
Micro-lens array	62
Microarrays	63
Photonic crystals	64
Lithium air battery	65
Ultrasonic coupling	66

applications and potential applications for hydrogels as described in the patent literature are listed in Table 13.4.

13.4 SUMMARY

Hydrogels are three-dimensional networks of hydrophilic polymers. Their applications rely on the ability to swell up on absorption of large quantities of water relative to their weight and to retain the absorbed water. Products containing hydrogels are quite common in everyday life, for example, in soft contact lenses, cosmetics, wound dressings and nappies. However, applications cover many areas including the treatment of soils in arid environments, surgical procedures, tissue engineering and groundwater barriers. The worldwide market for hydrogels is estimated to be £2.5 billion.

REFERENCES

1. His Honour Judge Birss QC, Samsung Electronics (UK) Limited v Apple Inc., Case number HC 11 C03050, Neutral

citation number [2012] EWHC 1882 (Pat). In the High Court of Justice, Chancery Division, Patents Court, 18th, 19th June 2012.

2. S. Kazakov, M. Kaholek and K. Levon, Nanogels and their production using liposomes as reactors, *United States Patent*, 7 943 067, 2011.

3. B. Thomas and K. Zhang, Perfluorocyclobutane crosslinked hydrogels, *United States Patent*, 8,017,107, 2011.

4. R. Mertens and O. Holler, Preparation of superabsorbent polymers, *United States Patent*, 7 612 016, 2009.

5. H. Omidian and J. G. Rocca, Superporous hydrogels for heavy-duty applications, *United States Patent*, 7 988 992, 2011.

6. S. J. Smith and E. J. Lind, Superabsorbent polymer having improved absorption rate and absorption under pressure, *European Patent*, 0 644 207 B, 2004.

7. O. Jordan, S. G. Kaderli and R. Gurny, Hybrid hydrogels, *International Patent Application*, WO 2014/032780, 2014.

8. J. Dentler, N. Herfert, H. Klotzsche, R. Schliwa and U. Stuven, Complete drying method for hydrogels, *United States Patent*, 6 641 064, 2003.

9. P. D. Unger and R. P. Rohrbach, Process for making low density hydrogel materials having high surface areas, *United States Patent*, 5 541 234, 1996.

10. Y.-C. Lai, D. V. Ruscio and P. L. Valint, Jr, Surface wettable silicone hydrogels, United States Patent, 5 387 632, 1995.

11. B. Kohler and M. Mager, Hydrogels of hydrophilic polyurethane (meth)acrylates, *United States Patent*, 7 947 863, 2011.

12. A. M. P. Wibaux, Hydrogel including modified cyclodextrin crosslinked with polyurethane prepolymer, *United States Patent Application*, 2008/0287604, 2008.

13. K. H. Guillen and A. Tezel, Polysaccharide and protein-polysaccharide cross-linked hydrogels for soft tissue augmentation, *United States Patent*, 8 691 279, 2014.

14. H. Omidian and J. Gutierrez-Rocca, Formation of strong superporous hydrogels, *United States Patent*, 7 056 957, 2006.

15. S.-C. Song, M.-R. Park and S.-M. Lee, Poly(organophosphazene) hydrogels for drug delivery, preparation method thereof and use thereof, *United States Patent*, 8 075 916, 2011.

16. S.-C. Song, T. Potta and S.-M. Lee, Phosphazene hydrogels with chemical cross-link, preparation method thereof and use thereof, *United States Patent*, 8 313 771, 2012.
17. W. E. Hennink and W. N. E. Van Dijk-Wolthuis, Biodegradable hydrogels, *International Patent Application*, WO 2006/071110, 2006.
18. A. W. Bosman, Fluorinated supramolecular polymers, *European Patent Application*, 2 457 940 A, 2012.
19. K. Park and H. Park, Super absorbent hydrogel foams, *United States Patent*, 5 750 585, 1998.
20. C. R. Mace, J. Barber, A. L. Sague, G. M. Whitesides and R. Cademartiri, Alginate hydrogel fibers and related materials, *United States Patent Application*, 2013/0316387, 2013.
21. H. B. Larman and F. Stellacci, Supramolecular nanostamping printing device, *United States Patent Application*, 2010/0256017, 2010.
22. D. H. Solomon, G. G. Qiao and A. Y. L. Kwok, Hydrogel preparation and process of manufacture thereof, *United States Patent Application*, 2007/0068816, 2007.
23. J. C. Salamone, Fluorine containing soft contact lens hydrogels, *United States Patent*, 5 684 059, 1997.
24. S. Rastogi, G. Friends and M. Nandu, Preferential release of an ophthalmic lens using a super-cooled fluid, *United States Patent Application*, 2007/0132120, 2007.
25. D. C. Turner, R. B. Steffen, C. Wildsmith and T. A. Matiacio, Method for manufacturing a contact lens, *United States Patent*, 7 442 710, 2008.
26. D. G. Vanderlaan, I. M. Nunez and M. Hargiss, Silicone hydrogel polymers, *European Patent*, 0 908 744 B, 2002.
27. C. L. Schultz, I. M. Nunez, D. L. Silor and M. I. Neil, Contact lens containing a leachable absorbed material, *United States Patent*, 5 723 131, 1998.
28. J. C. Phelan, Curable colored inks for making colored silicone hydrogel lenses, *United States Patent*, 7 550 519, 2009.
29. D. C. Vanderlaan, D. C. Tumer, M. V. Hargiss, A. C. Maiden, R. N. Love, J. D. Ford, F. F. Molock, R. B. Steffen, A. Alli, J. B. Enns and K. P. Mccabe, Soft contact lenses, *United States Patent Application*, 2004/0186248, 2004.
30. G.-H. Ho, T.-H. Yang and J. Yang, Moisturizers comprising one or more of gamma-polyglutamic acid (gamma-PGA, H

form), gamma polyglutamates and gamma-polyglutamate hydrogels for use in cosmetic or personal care products, *United States Patent Application*, 2009/0110705, 2009.

31. M. Dalko, Cosmetic processes with glucosamine-based hydrogels, *International Patent Application*, 2013/092872, 2013.

32. M. Lai, J. Vidwans and Q. Wu, Soap bar containing hydrogel phase particles, *United States Patent*, 8 618 035, 2013.

33. G. Palumbo and G. Carlucci, Layered absorbent structure, *United States Patent*, 5 482 761, 1996.

34. D. McGhee, Y. H. Huang, S. B. Earhart and W. R. Fiehler, Hydrogel wound dressing and the method of making and using the same, *United States Patent*, 6 861 067, 2005.

35. P. J. Davis, A. J. Austin and J. Jezek, Wound dressings comprising hydrated hydrogels and enzymes, *United States Patent Application*, 2010/0183701, 2010.

36. D. Silcock and A. J. Kirkwood, Absorbent wound dressing containing a hydrogel layer, *United States Patent*, 8 058 499, 2011.

37. D. Addison, Layered materials for use as wound dressings, *United States Patent*, 8 097 272, 2012.

38. G. Palumbo and J. Preuschen, Hydrogel pressure sensitive adhesives for use as wound dressings, *European Patent Application*, 1 051 982 A, 2000.

39. W. G. Meathrel, M. Saleem and S. A. Binks, Hydrogel adhesive for attaching medical device to patient, *United States Patent*, 5 665 477, 1997.

40. S. A. Goldman and W. E. Huhn, Polymerized hydrogel adhesives with low levels of monomer units in salt form, *United States Patent Application*, 2004/0115251, 2004.

41. H. S. Munro, Bioadhesive compositions, *United States Patent*, 6 683 120, 2004.

42. A. Huttermann, Watering method and system for soil comprising hydrogel, *United States Patent Application*, 2010/0050506, 2010.

43. M. H. Savich, Superabsorbent polymer root dip, *United States Patent*, 7 607 259, 2009.

44. T. M. Schultz, A. H. Finnel and G. L. Collins, Seed coating hydrogels, *United States Patent Application*, 2014/0100111, 2014.

45. P. Palmer, T. Schreck, L. Hamel, S. Tzannis and A. Poutiatine, Small volume oral transmucosal dosage forms containing sufentanil for treatment of pain, *United States Patent*, 8 535 714, 2013.
46. P. Ashton and J. Chen, Polymeric gel delivery system for pharmaceuticals, *United States Patent Application*, 2012/0195934, 2012.
47. E. Donati and I. Rapaport, Plaster for topical use containing heparin and diclofenac, *United States Patent*, 7 799 338, 2010.
48. H. P. Benecke, B. Vijayendran and K. B. Spahr, Absorbent protein meal base hydrogels, *European Patent*, 2 249 877 B, 2012.
49. R. Wicker, F. Medina, K. Arcaute, L. Ochoa, C. Elkins and B. Mann, Hydrogel constructs using stereolithography, *United States Patent*, 8 197 743, 2012.
50. N. Bartetzko. Printable hydrogels for biosensors, *United States Patent Application*, 2010/0166607, published on 1 July 2010.
51. K. Yasuda, Y. Osada, J. P. Gong and N. Kitamura, Method for bone-filling cartilage tissue for inducing regeneration of the cartilage, *United States Patent Application*, 2011/0257763, 2011.
52. W. Skalla and N. Mast, Mesh implant, *United States Patent*, 8 470 355, 2013.
53. C. Le Visage and D. Letourneur, Method for preparing porous scaffold for tissue engineering, cell culture and cell delivery, *European Patent*, 2 203 194 B, 2013.
54. M. Wei and Z. Xia, Biomimetic scaffold for bone regeneration, *United States Patent Application*, 2013/0251762, 2013.
55. Y. Li and J. Zhou, Hydrogel for use in downhole seal applications, *United States Patent Application*, 2006/0047028, 2006.
56. M. D. Lockhart, Groundwater isolation barriers for mining and other subsurface operations, *United States Patent*, 8 381 814, 2013.
57. L. P. Requejo, Fragranced hydrogel air freshener kits, *United States Patent Application*, 2002/0041860, 2002.
58. T. McGee and R. P. Sgaramela, Substrate care product, *United States Patent Application*, 2008/0234172, 2008.
59. L. Pan, D. Zhai, Y. Shi and H. Qiu, Conductive polymer synthesis method thereof, and electroactive electrode

covered with said conductive polymer, *United States Patent Application*, 2013/0171338, 2013.

60. K.-B. Lee and S. Shah, Stem cell differentiation using novel light-responsive hydrogels, *United States Patent Application*, 2012/0149781, 2012.

61. D. W. Goupil, H. Chaouk, T. Holland, B. T. Asfaw, S. D. Goodrich and L. Latini, Hydrogel biomedical articles, *United States Patent*, 7 070 809, 2006.

62. J. Aizenberg, M. Megens and S. Yang, Tunable micro-lens arrays, *United States Patent*, 7 106 519, 2006.

63. P. Tsinberg, Microarrays utilizing hydrogels, *United States Patent*, 7 595 157, 2009.

64. V. L. Alexeev, Hydrogel photonic crystals that can be de-hydrated and re-hydrated, *United States Patent Application*, 2008/0157035, 2008.

65. S. J. Visco, L. C. De Jonghe, Y. S. Nimon, A. Petrov and K. Pridatko, Hydrogels for aqueous lithium/air battery cells, *United States Patent*, 8 389 147, 2013.

66. G. K. Lewis, Jr., J. L. Guarino and B. Guffey, Hydrogel ultra-sound coupling device, *United States Patent Application*, 2013/0144193, 2013.

CHAPTER 14

Superhydrophobic Materials

14.1 INTRODUCTION

Anyone who has purchased a raincoat may have noticed that the label often refers to a waterproofing treatment applied to the fabric so that water does not penetrate the fabric. Water that is spilled onto a wooden table can spread out as a film or break up into droplets, depending on whether the table has been polished. Brickwork in a house can be treated with proprietary coatings to prevent ingress of rain into the brickwork. The leaf of the lotus plant exhibits water-repellency and self-cleaning properties.[1] Although the plant often grows in muddy rivers and lakes, the leaves remain clean and non-wettable. The leaves and flowers have hydrophobic surfaces and when they come into contact with water, the water droplets contract into spherical beads which roll off the surface, sweeping away particles of dirt they encounter. Thus, the leaf is self-cleaning. The leaves of many plants and the wings of insects and birds are able to repel water. These examples involve the surfaces of objects and people encounter surfaces every day. Superhydrophobicity is a physical property of a surface whereby the surface is extremely difficult to wet.[1] Hence superhydrophobic surfaces (or super water repellent surfaces) refer to surfaces that have very high repellency or extremely low affinity for water and are the subject of this chapter.[2]

Exploring Materials through Patent Information
By David Segal
© David Segal, 2015
Published by the Royal Society of Chemistry, www.rsc.org

14.2 SUPERHYDROPHOBIC MATERIALS

14.2.1 Definition in Terms of Contact Angle

Contact angles were mentioned in Chapter 9 in relation to hydrophobic aerogels. Figure 14.1 illustrates an idealised view of liquid droplets on a surface.[3]

On an extremely hydrophilic surface, a water droplet will completely spread over the surface and exhibit a contact angle, θ_c, of approximately 0°. This situation arises for surfaces that absorb water or have a high affinity for water. Many hydrophilic surfaces have contact angles from around 10° to 30°.[4] Hydrophobic surfaces are associated with larger contact angles, around 70° to 90° and above. Some hydrophobic surfaces such as Teflon™ exhibit a contact angle of between 120° to 130°. The lotus leaf has a contact angle in contact with water of about 160°. A superhydrophobic surface is considered to exhibit a contact angle greater than about 140°. A non-wetting surface has a contact angle of 180°. It is considered that the three-dimensional structure of the lotus leaf has wax crystals that self-organise to provide roughness on the nanometre or micrometre scale. Protuberances on the hydrophobic surface reduce the effective surface contact area with water, increase the surface roughness and hence prevent adhesion and spreading of the water over the leaf.

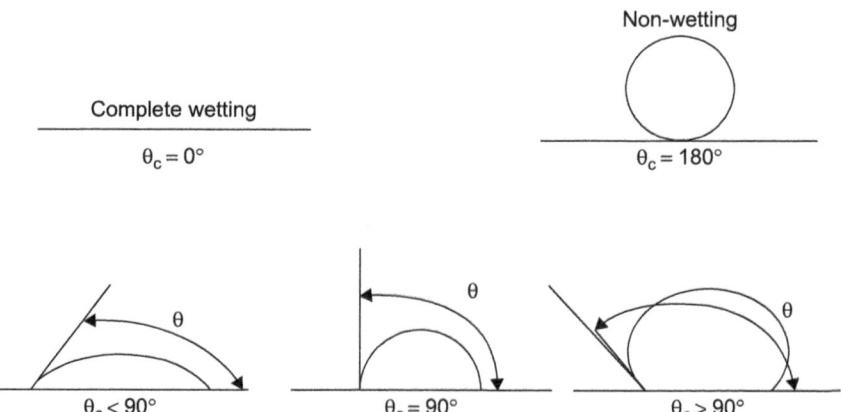

Figure 14.1 Contact angles at a water–solid interface.[3]

14.2.2 Surface Roughness and Reduced Surface Energy

The hydrophobic nature of surfaces is increased by increasing the contact angle at the air-water-solid interface. This can be achieved through increasing the surface roughness and lowering the surface energy by applying a liquid with a low surface tension such as a fluorine-based material. Substrate surfaces that are smooth at the molecular level have been made hydrophobic by a variety of methods including deposition of a layer of lipid molecules or fluorocarbons with polar head groups or by a specific chemical reaction.[5] For example, treatment of a thin gold layer that has been deposited on a substrate surface with an alkyl thiol. These methods allowed the contact angle for a water droplet on a smooth surface to be increased to between 100 and 120 degrees.

14.2.3 Hierarchical Structures

Another general approach to produce superhydrophobic materials is to engineer surfaces with a hierarchical structure,[6] that is, with features of both microscale and nanoscale dimensions. Hierarchical structures appear smooth to the naked eye at the macroscale but are not smooth at smaller scales, namely at the scale of micrometres (microscale) or nanoscale (nanometre). Such surfaces can be considered to be biomimetic structures that are similar to the surface structure of natural organisms such as the lotus plant.

An example of the hierarchical approach concerns ophthalmic lenses,[7] in particular spectacle glasses. Hydrophobic surfaces to prevent water droplets from remaining stuck on the glass lenses have been derived from fluorinated compounds such as fluorosilanes and fluorosilazanes and contact angles with water of 110–120° can be obtained. The hierarchical approach involves developing a nanotextured periodic array on the surface of the lenses. A layer of poly(methyl methacrylate) 150 nm in thickness was deposited by spin coating onto a silicon wafer, placed in an oven for 30 minutes at 453 K and cured by electron beam irradiation using a masking device. The irradiated layer is developed in a solution of methylisobutylketone and isopropanol, after which a chromium layer was deposited by vacuum deposition. The poly(methyl methacrylate) layer is removed by treatment with acetone so that chromium deposited on the silicon remains intact and the chromium above the resin is removed with the resin. The silicon is then subjected to reactive ion etching to obtain an etching depth of 350 nm and the chromium layer was then chemically removed. This process produced an array of vertical pillars with thicknesses between 100 nm and 400 nm on the silicon. A layer of hydrophobic material with a thickness of 2 nm was then deposited onto the array of vertical pillars on the silicon by evaporation and the array had a contact angle of 160°, characteristic of a superhydrophobic material, compared to 121° for the silicon substrate.

Nanomaterials are materials composed of nanoscale particles and include quantum dots, ceramic nanoparticles, metallic nanoparticles and biological nanomaterials.[6] The latter include viruses that consist generally of a protein coat surrounding viral genetic material and in some cases a lipid envelope surrounding the protein coat. Biomimetic hierarchical surfaces based on the Tobacco mosaic virus (TMV) have been developed.[6] This is a plant virus measuring 300 nm in length and 18 nm in diameter and can be genetically engineered to enhance binding properties. Engineered functional groups at predetermined locations enable self-assembly and directed patterning over a range of materials including metals, ceramics and polymers. The following example illustrates the type of hierarchical structure that can be obtained with TMV. A photoresist was spin-coated onto a silicon wafer to a thickness of 15 μm, after which the wafer was exposed to produce micropost arrays. The wafer was diced into individual 2 cm×2 cm dies that were placed in a phosphate buffer solution containing TMV that self-assembled on the exposed silicon and resist surfaces. Exposed cysteine residues on the virus were activated with a solution of a palladium catalyst and then the substrates were coated with nickel in an electroless plating solution. After metallisation, the surfaces were functionalised through atomic layer deposition of alumina followed by vapour phase deposition of silane to achieve superhydrophobic properties with contact angles greater than 170°. The virus is fully encased and the final nanostructure is a three-dimensional scaffolding of TMV coated with metal. This fabrication process could produce arrays of protuberances attached vertically to the surface of the wafer.

A further example illustrates how a superhydrophobic polymer or metal surface can be obtained by using a textured surface with materials at three different size scales.[8] Polyvinylidene fluoride (PVDF) beads contain particles with an average diameter of 130 nm (nanoscale particles) while the beads have diameters in the range 0.5–50 μm. A layer of beads was spread as a powder onto a sheet of polyethylene and exposed to hot pressing with a riffled stamp. The sheet softens and traps individual PVDF particles and aggregates of PVDF beads. The stamp produced indentations or channels with a depth of 20 μm and separation of 100 μm. This textured structure exhibited a contact angle of 160°.

Reviews of methods for the preparation of superhydrophobic materials are described in the patent literature.[9]

14.3 APPLICATIONS FOR SUPERHYDROPHOBIC MATERIALS

14.3.1 The Range of Applications

Examples of the potential range of applications[10–12] for superhydrophobic materials are shown in Table 14.1.

14.3.2 De-icing

Satellite transmissions for commercial television are broadcast[13] from satellites in geosynchronous orbits to satellite antenna systems designed to receive the signals. Transmissions are delivered to dish-shaped antenna that are usually about eighteen inches in diameter. The antenna dish concentrates and reflects satellite microwave signals that strike the antenna dish back to a

Table 14.1 The range of applications for superhydrophobic materials.[10–12]

Applications
Anti-biofouling paints for boats
Anti-sticking of snow for antennas, windows, wind turbines aircraft wings
Corrosion prevention
Prevention of icing on high-voltage cables and flashovers
Self-cleaning components in xerography
Self-cleaning windscreens
Stain resistant textiles

focal point that is in front of the antenna dish. A feed horn that is positioned on a support arm at the focal point directs the microwave signals to a signal converter that converts the microwave signals into electrical signals. The electrical signals are then provided to a satellite receiver that translates the electrical signals into a television picture and sound. Weather conditions such as rain, ice, dew, wind or snow can cause the signal to fade, resulting in poor reception. Superhydrophobic materials have been applied to reflective parts of the antenna with mixed results on the signal reception. However, application of superhydrophobic materials to transmissive surfaces, in particular to the exterior surface of the feed horn cover that microwaves pass through gave a more consistent satellite signal.

Patterned hydrophobic surfaces can repel water droplets contacting a substrate and reduce the contact time for nucleation of ice and its adhesion to the surface.[14] The surface has raised structures obtained by photolithography and the structures can consist of arrays of cylinders, honeycombs, boxes or bricks. The height of the structures extends to a maximum of 1000 μm. A hydrophobic layer of a fluorinated compound, about 2 nm in thickness, was deposited by plasma assisted chemical vapour deposition Non-wetting water droplets froze after wetting droplets on the patterned surface. The build-up of ice on aircraft surfaces can be a potential risk to the passengers and crew.[15] A nanoporous titanium dioxide layer about 1 nm to 10 nm in thickness and containing nanotubes was obtained on anodising titanium. A hydrophobic coating of a fluorocarbon was deposited onto the rough surface in a plasma process to produce a self-cleaning superhydrophobic substrate.

14.3.3 Bandages

Breathable bandages can, over a period of time, absorb water, particularly when washing adjacent regions of skin so that they need to be replaced.[16] Figure 14.2 is the cross-section of a bandage that contains superhydrophobic particles.

A breathable bandage, namely one that is permeable to water vapour and gases is placed over a wound dressing so that water vapour can pass through the bandage away from the skin. However, a layer of superhydrophobic material is embedded in

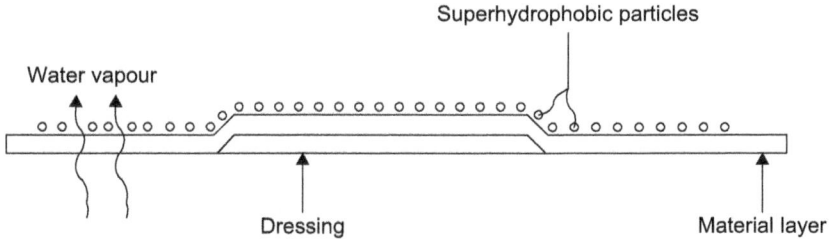

Superhydrophobic particles

Water vapour

Dressing

Material layer

Figure 14.2 Schematic diagram of a breathable bandage with super-hydrophobic particles.[16]

an outer material layer that is not superhydrophobic and prevents the transport of water vapour through the bandage towards the skin. Superhydrophobic material was based on a silica powder with a particle size in the range 0.5 μm to about 7 μm and derived from a phase separated glass. A monolayer of tridecafluoro-1,1,2,2-tetrahydrooctyl trichlorosilane was deposited onto the silica particles from solution, rendering them superhydrophobic.

14.3.4 Membranes

Membranes have a variety of applications including as separators in lithium-ion batteries, filtration and desalination and it is important to avoid fouling of the membranes in practice in order to maintain efficient processes. Polyvinylidene fluoride (PVDF) membranes have attractive physical properties, namely hydrophobicity, heat resistance, flexibility and chemical resistance.[17] PVDF was dissolved in a solvent and spread out a solid support to form a film. The film was first immersed in an alcohol such as ethanol and then immersed in a bath of water and dried. The resulting membranes had a contact angle of around 150° and a hierarchical structure. Hence interconnected crystalline nodules with sizes between 5 μm and 12 μm and an inter-nodular porosity less than 5 μm were present with an intra-nodular porosity of a few hundred nanometres. This hierarchical sponge-like structure was considered to be the cause of superhydrophobicity as the two levels of organisation, inter-nodular and intra-nodular porosity can trap air and prevent close contact between the surface and water.

Electrospinning has been used to prepare polymer fibres, in which a solution is extruded from a spinerette under the influence of an electric field. A superhydrophobic membrane has been prepared by electrospinning a solution of polyvinylidene fluoride containing a water-repellent additive.[18] An example of such an additive is poly (2,2,2-trifluoroethyl 2-fluoroacrylate). Phase separation in block copolymers was described in Chapter 8 with specific reference to applications in the semiconductor industry. These materials have found application in the preparation of superhydrophobic fibres.[19] A poly(styrene-co-dimethylsiloxane) diblock copolymer was dissolved in a mixture of tetrahydrofuran and dimethylformamide and electrospun to fibres with diameters in the range 150 to 400 nm. Phase separation occurred in the fibres that appeared to consist of cylinders of polydimethylsiloxane with a diameter of about 20 nm dispersed in a polystyrene matrix. This hierarchical structure contributed to the superhydrophobic behaviour of the fibres that had a water contact angle greater than 150°. Electrospinning was combined with initiated chemical vapour deposition (iCVD) to produce superhydrophobic fibres, for example, polyester fibres.[20] For example, polycaprolactone was electrospun to mats and coated with a polymer by iCVD. In this process, a monomer and initiator are fed into a reactor and the initiator is broken down by heat or ultraviolet light to form free radicals. The latter polymerise the monomer that results in a coating of polymer on the fibrous mat. An example of a polymer produced in iCVD is poly(perfluoroalkyl ethyl methacrylate). The hierarchical structure of the electrospun mat that contained both nano-scale fibres and micron-scale beads and interfibrillar distances was considered to contribute to the superhydrophobic nature of the fibres because air trapped in apertures between the fibres and the polymer layer prevented water from sinking into cavities in the mat.

14.3.5 Cosmetics

Hydrophobic materials are used in cosmetics, for example in mascara, lipstick, lip gloss, nail polish and sunscreens to impart waterproof or water-resistant compositions, but conventional materials are not superhydrophobic, that is, the water contact angle is less than 140°.[4,21] This is because they do not exhibit surface roughness at the nanoscale or microscale.[21] Cosmetic films that have superhydrophobic properties are desirable as they can have improved water repellency, self-cleaning properties and long-wear properties. Superhydrophobic materials have been developed for cosmetics. For example, in the case of mascara as with other cosmetics the compositions are proprietary. However, cosmetic compositions usually contain a hydrophobic film former, for example, ethylene/propylene/styrene and butylene/ethylene/styrene copolymer and a hydrophobically modified iron oxide pigment in the case of mascara;[21] for example, a perfluoroalkylsilane-treated iron oxide pigment such as perfluorooctyl triethoxysilane-treated iron oxide. Other components of mascara can include keratin fibres for eyelashes. The iron oxide pigment can be black, red, yellow, brown, orange or blue with particle sizes between 0.05 μm and 20 μm.[21] It is considered that particulates provide nanoscale (1 nm–1000 nm) or microscale (1 μm–200 μm) surface roughness which imparts superhydrophobicity in which water droplets sit on surface protuberances, thus minimising surface adhesion. The silane-treated iron oxide can be combined with carbon black to increase the contact angle to around 150°, the region of superhydrophobicity.

14.3.6 Medical devices

Stents are generally tubular devices that are used to support a segment of a blood vessel.[22] During insertion, the stent maintains an expanded state but in some stents there is a possibility that the expanded configuration may change. There is evidence that superhydrophobic regions will spontaneously bind together in the presence of water and if these regions are on a stent then possible changes in the shape of the stent are reduced. Superhydrophobic regions can be obtained by texturing a fluorocarbon layer by, for example, plasma etching or by depositing a fluorocarbon layer over a textured region of the substrate.

14.3.7 Treatment of Oil Spills

Oil spills endanger the environment and can be costly to remove. Superhydrophobic materials have potential for the treatment of oil spills.[23] A porous material such as cellulose or a foamed polymer is soaked in a dispersion of graphene sheets in alcohol, where the sheets have a thickness of 0.5–5.0 nm and length of about 0.15–10.0 μm. The porous material is dried at 373 K in a vacuum chamber so that rough surfaces are formed, as graphene sheets adhere to the skeletal structure of the porous material. The porous material is then soaked in a solution of poly-dimethylsiloxane that acts as an adhesive for the graphene sheets and dried to form a composite. Water droplets do not penetrate the composite whereas motor oil, for example does penetrate the composite. The surface of the composite has a contact angle of around 160° that represents a superhydrophobic material. The composite can be reused after removal of absorbed oil by, for example, distillation or centrifugation.

14.3.8   Additional Applications

Earlier chapters referred to specific classes of materials such as ionic liquids and flame retardants. However, superhydro-phobicity is a surface property of materials and is associated with texturing of the surface to enhance the surface roughness. Thus, materials that are not naturally superhydrophobic can be made to have this property, as noticed with spectacle lenses. Some additional references in the patent literature are shown in Table 14.2 to illustrate the scope of activity in this area.

Table 14.2   Additional examples of superhydrophobic materials.

Application	Reference
Superhydrophobic aerogel	24
Superhydrophobic powders	25
Superhydrophobic anodized metals	26
Superhydrophobic powders	27
Superhydrophobic gypsum boards	28
Superhydrophobic transparent surfaces	29
Plasma-derived coatings	30
Plasma derived coatings	31
Hierarchical structures	32
Hierarchical structures	33

14.4 SUMMARY

Superhydrophobic materials have contact angles at the solid/air/ water interface of 140° or greater. They have low affinity for water and are extremely difficult to wet. They can be prepared by surface texturing to introduce a hierarchical structure with features at both the microscale and nanoscale combined with a layer of a fluorinated compounds. The materials are self-cleaning and have many applications and potential applications including cosmetics, bandages, de-icing systems, membranes, spectacle lenses, medical devices and removal of oil spills.

REFERENCES

1. J.-K. Lee, Y. Lee and K.-Y. Ju, Superhydrophobic polymer fabrication, *United States Patent*, 8 455 084, 2013.
2. A. Laukkanen, J.-E. Teirfolk, O. Ikkala, R. Ras and H. Mertaniemi, Hydrophobic coating and a method for producing hydrophobic surface, *International Patent Application*, 2012/152997, 2012.
3. T. Ono, Method of making a golf ball with a super-hydrophobic surface, *United States Patent Application*, 2013/0287967, 2013.
4. R. A. Ranade, J. R. Glynn, M. S. Garrison, S. Martin and P. Maitra, Cosmetic compositions for imparting super-hydrophobic films, *United States Patent Application*, 2010/0266648, 2010.
5. O. P. Werner, L.-E. R. Wagberg, C. Quan, C. K. Turner and J.-C. Eriksson, Method to prepare superhydrophobic surfaces on solid bodies by rapid expansion solutions, *United States Patent*, 8 722 143, 2014.
6. E. N. Wang, M. McCarthy, R. Enright, J. N. Culver, K. Gerasopoulos and R. Ghodssi, Superhydrophobic surfaces, *United States Patent Application*, 2013/0059123, 2013.
7. M. Reyssat, Y. Chen, A. Pepin, D. Quere, C. Biver and L. Vagharchakian, Article having a nanotextured surface with superhydrophobic properties, *United States Patent*, 8 298 649, 2012.
8. E. Bormashenko, Y. Bormashenko, G. Vaiman and T. Stein, Superhydrophobic nanotextured polymer and metal

surfaces, *United States Patent Application*, 2010/0021692, 2010.

9. J. T. Simpson, Superhydrophilic and superhydrophobic powder coated fabric, *United States Patent Application*, 2009/0042469, 2009.

10. D. R. Strauss, Self-cleaning superhydrophobic surface, *United States Patent*, 8 580 371, 2013.

11. C. J. Greyling, Superhydrophobic coatings and methods of preparation, *International Patent Application*, 2013/042052, 2013.

12. Y. Qi, N.-X. Hu, Q. Zhang, G. Song and S. J. Gardner, Superhydrophobic nano-fabrics and coatings, *European Patent Application*, 2 210 921 A, 2010.

13. L. D. King, Antenna systems for reliable satellite television reception in moisture conditions, *United States Patent*, 7 342 551, 2008.

14. B. Hatton, L. Mishchenko, J. Alzenberg, T. Krupenkin and J. A. Taylor, Patterned superhydrophobic surfaces to reduce ice formation, adhesion and accretion, *United States Patent Application*, 2013/0227972, 2013.

15. D. R. Strauss, Self-cleaning superhydrophobic surface, *United States Patent*, 8 580 371, 2013.

16. J. T. Simpson and B. R. D'Urso, Super-hydrophobic bandages and method of making the same, *United States Patent*, 8 193 406, 2012.

17. A. Deratani, D. Quemener, D. Booyer, C. Pochat-Bohatier, C.-L. Li, J.-Y. Lai and D.-M. Wang, PVDF membranes having a superhydrophobic surface, *United States Patent Application*, 2013/0306560, 2013.

18. S. O. Kim, J. W. Na, K. M. Jeon and J. S. Ryoo, Super-hydrophobic membrane and method of manufacturing the same, *United States Patent Application*, 2014/0065906, 2014.

19. G. C. Rutledge, R. M. Hill, J. L. Lowery, M. Ma and S. Fridrikh, Superhydrophobic fibers and methods of preparation and use thereof, *United States Patent*, 8 574 713, 2013.

20. K. K. Gleason, G. C. Rutledge, M. Gupta, M. Ma and Y. Mao, Superhydrophobic fibers produced by electrospinning and chemical vapour deposition, *United States Patent*, 7 651 760, 2010.

21. R. A. Ranade and M. S. Garrison, Cosmetic compositions for imparting superhydrophobic films, *United States Patent Application*, 2011/0008401, 2011.
22. M. Edin, Medical devices having superhydrophobic surfaces, *United States Patent*, 8 043 359, 2011.
23. N.-H. Tai, Superhydrophobic and superoleophilic composite, *United States Patent Application*, 2014/0057782, 2014.
24. D. J. Kissel and C. J. Brinker, A superhydrophobic aerogel that does not require per-fluoro compounds or contain any fluorine, *European Patent*, 2 294 152 B, 2012.
25. J. T. Simpson and S. R. Hunter, Method of making superhydrophobic/superoleophilic paints, epoxies and composites, *United States Patent Application*, 2014/0090578, 2014.
26. C. N. Barbier, J. T. Simpson, B. R. D'Urso and E. Jenner, Superhydrophobic anodized metals and making the same, *United States Patent Application*, 2014/0110263, 2014.
27. R.-H. Jin and J.-J. Yuan, Superhydrophobic powders, structure with superhydrophobic surface, and processes for producing these, *United States Patent Application*, 2011/0195181, 2011.
28. J.-P. Boisvert and G. Hedman, Superhydrophobic gypsum boards and process for making same, *United States Patent Application*, 2008/0245012, 2008.
29. A. M. Lyons and Q. Xu, Polymer having optically transparent superhydrophobic surface, *United States Patent Application*, 2014/0106127, 2014.
30. B.-K. Kang, Surface coating method for hydrophobic and superhydrophobic treatment in atmospheric pressure, *United States Patent Application*, 2010/0221452, 2010.
31. R. D'Agostino, I. Corzani, P. Favia, R. Lamendola and G. Palumbo, Modulated plasma glow discharge treatments for making superhydrophobic substrates, *European Patent*, 1 112 391 B, 2004.
32. B. Bhushan, Y. C. Jung and M. Nosonovsky, Hierarchical structures for superhydrophobic surfaces and methods of making, *United States Patent*, 8 137 751, 2012.
33. J. T. Simpson, C. A. Blue and J. O. Kiggans, Jr, Article coated with flash bonded superhydrophobic particles, *United States Patent*, 7 754 279, 2010.

Subject Index